SMALL-SCALE TEXTILES

SERICULTURE AND SILK PRODUCTION

Other books in this series:

Spinning

Dyeing and Printing

Yarn Preparation

Fabric Manufacture

Medical and Hygiene Textile Production

Plant Fibre Pre-processing

SMALL-SCALE TEXTILES

SERICULTURE AND SILK PRODUCTION

A handbook

Prabha Shekar and Martin Hardingham

Intermediate Technology Publications 1995

Practical Action Publishing Ltd
27a Albert Street, Rugby, CV21 2SG, Warwickshire, UK
www.practicalactionpublishing.org

© Intermediate Technology Publications 1995

First published 1995\Digitised 2013

ISBN 10: 1 85339 317 7
ISBN 13: 9781853393174
ISBN Library Ebook: 9781780443249
Book DOI: http://dx.doi.org/10.3362/9781780443249

A catalogue record for this book is available from the British Library.

Since 1974, Practical Action Publishing has published and disseminated books and
information in support of international development work throughout the world.
Practical Action Publishing is a trading name of Practical Action Publishing Ltd
(Company Reg. No. 1159018), the wholly owned publishing company of Practical
Action. Practical Action Publishing trades only in support of its parent charity
objectives and any profits are covenanted back to Practical Action (Charity Reg. No.
247257, Group VAT Registration No. 880 9924 76).

CONTENTS

ACKNOWLEDGEMENTS

It would be impossible to mention all the people who have helped in the research and production of this handbook. We are grateful to our many friends at the Central Silk Board (SB), the Central Silk Technological Research Institute (CSTRI) in Bangalore, South India, the Central Sericultural Research and Training Institute (CSR+TI), International Centre for Training and Research in Tropical Sericulture in Mysore, South India, and to the Department of Sericulture, Government of Karnataka, India. All of these institutions have generously provided much of the statistical and technical information for this handbook.

Our thanks are due also to Ms Maya Sitaram, Mr Mahadeva Gowda Patil and Mr Linganna Gowda for their assistance.

We are particularly grateful for the excellent illustrations drawn especially for this handbook by Sue Sharples and Ethan Danielson.

Prabha Shekar
Martin Hardingham

FOREWORD

This handbook is one of a series dealing with small-scale textile production, from raw materials to finished products. Each handbook sets out to give some of the options available to existing or potential producers, where the aims could be to create employment or to sustain existing textile production. The ultimate goal is to generate incomes for the rural poor in developing countries.

Needless to say, this slim volume does not pretend to be comprehensive. Sericulture and silk production are vast subjects. This handbook is primarily technical and is intended as an introduction which may stimulate further enquiry.

I am pleased to have had the opportunity of working with Prabha Shekar on this handbook, which has been sponsored by Intermediate Technology, UK. We hope that this handbook will help to identify some of the most appropriate solutions to particular development problems. The series of small-scale textile handbooks forms part of the process of identifying the need, recognizing the problems, and developing strategies to alleviate the crisis of unemployment and underemployment in the South.

Intermediate Technology, UK, offers consultancy and technical enquiry services.

Martin Hardingham
Intermediate Technology, UK

PREFACE

This handbook sets out to provide basic information about sericulture and silk production and deals only with mulberry silk. The sericultural practices referred to in this book are mainly those followed in India and are suitable for countries with similar agro-climatic conditions.

The involvement of Intermediate Technology, UK, in the Indian silk industry goes back to 1988 when Economic Development Associates, New Delhi, were commissioned to study the techno-economic issues involved in the production of silk in India. This study by Sanjay Sinha, published as *The Development of Indian Silk*, paved the way for a programme of work by Intermediate Technology to improve the silk-reeling device known as a *charaka*, which is widely used in south India.

The establishment of a project in Bangalore, south India, created direct links with the Central Silk Board of India. Numerous people assisted with the project, but it was Mr V. Balasubramanian, Member Secretary of the Central Silk Board (CSB), who was its guiding light.

A sericulture and silk-production project was established by the Intermediate Technology Zimbabwe office in 1992, which led to further interest from other African countries in silk production as a potential employment and income-generating activity. The need for a basic technical sericulture and silk-production handbook became increasingly obvious.

Today, although there is a great demand for silk, it is produced in limited quantities and accounts for only about 0.2 per cent of the total world production of textile fibres. This modest amount of silk is produced by 40 countries situated in temperate, subtropical and tropical belts. The world demand for pure silk continues to increase, but the supply is never met.

There are about 150 different types of silk in the world, of which four are widely known: *mulberry, tussah (tasar), eri* and *muga*. Mulberry silkworms are farmed while all other types are either semi-domesticated or wild. India is the only country in which all four varieties of silk are produced.

Sericulture and silk production, as rural cottage industries, have the potential of being able to create employment and incomes for rural people in many developing countries.

Prabha Shekar
Seri Tech Associates, Bangalore, South India

1. INTRODUCTION TO SILK AND SERICULTURE

SILK

Silk is a continuous-filament protein fibre which forms the cocoon made by the silkmoth larvae. The word 'sericulture' is derived from the Latin *sericum* which means 'the cultivation of silk'.

Silk is very thin but strong, beautiful and durable. Silk is unique because it is the only natural continuous filament used in the manufacture of textiles.

Historical evidence shows that the Chinese were the first to develop silk and to reel it from the cocoon more than 4500 years ago. For nearly 3000 years, silk production in China was a closely guarded secret, but gradually the knowledge spread eastwards to Japan and to North and South America, and westwards to India, Persia and Europe.

The ancient silk road started in what is now known as Xiang in the Shaan Si province of China. It crossed the deserts of central Asia and went on to Antioch and Tyre. The last lap of the silk route was on water, which took the silk-laiden boats into many Mediterranean ports. The most advanced civilizations of that period reigned at each end of the silk route. It was silk which formed the cultural bridge between the two.

SERICULTURE

Sericulture is the cultivation of silkworms for the production of cocoons from which silk is unwound to produce a textile thread. *Moriculture* is the cultivation of mulberry bushes.

Once mulberry bush plantations have been established, the sericulture chain begins with the selection of healthy moths for breeding. These moths produce eggs which are distributed to farmers, who hatch the silkworms and feed them continuously until they are ready to produce their cocoons. The cocoons are supplied to reelers, where they may be hand reeled on traditional machines or reeled on modern, automatic machines.

This handbook is concerned with cultivated silk and does not examine the production, harvesting or processing of silk found in the bush or in the forests of many tropical countries. Although there are no fundamental differences between sericulture in one country and another, the production processes vary.

The silkmoth

The silkmoth, of which there are many varieties, belongs to the lepidoptera group of insects. All lepidopteran larvae, which we call silkworms, feed on leaves or other vegetation to produce silk in the form of a cocoon. The *Bombyx mori* larvae feed on mulberry leaves and produce strong, long, lustrous filament silk used in textiles.

The eggs of the *Bombyx mori* silkmoth are as small as a pinhead. It takes from ten days to ten months for an egg to hatch, depending on the racial characteristics of the moth and the prevailing environmental conditions. The newly hatched black caterpillar, measuring 2 to 3mm in length, with hair all over its body, starts searching for food immediately after it has hatched. The silkworm feeds continuously for about 25 to 28 days reaching between 100 and 120mm in length. Periodically it rests from eating, appearing to sleep for about a day. On waking up, it wriggles out of its skin that has become too tight, a process referred to as moulting, and with its new stronger jaws it starts eating again. The silkworm undergoes several *moults*, normally casting its old skin four times in its larval stage. About eight to ten days after the final moult, the larva finds a place to make its cocoon.

Male   Female

Illustration 1 Male and female **Bombyx mori** *silkmoths (actual size)*

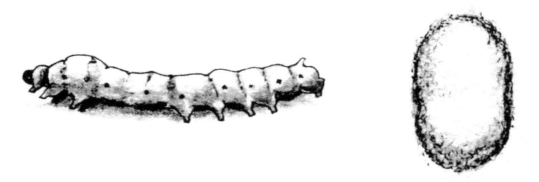

Illustration 2 Fully grown **Bombyx mori** *silkworm and its cocoon (actual size)*

While it is resting, major changes take place inside the cocoon as the larva shrinks and transforms itself into an immobile *pupa* or *chrysalis*. Within a fortnight the pupa changes into a moth. In order to get out of the cocoon, the moth emits an enzyme which softens the cocoon wall through which it pushes a hole.

Since the mulberry silkmoth has gone through a long period of domestication over several thousands of years, the silk moth has lost its ability to fly and now depends solely on man for its survival. The newly born adult moth is covered with scales, giving it an off-white shade. The female is fat and docile, unlike its mate which is active. It is the male which goes in pursuit of the female, attracted by her scent produced by her *pheromones*. After several hours of mating, the female starts laying eggs, which are normally 350 to 600 in number. The male and the female moths only live for two or three days. It is important to realize that in order to obtain a continuous filament of silk, this life cycle must be broken at the cocoon stage, before the pupa changes into a moth.

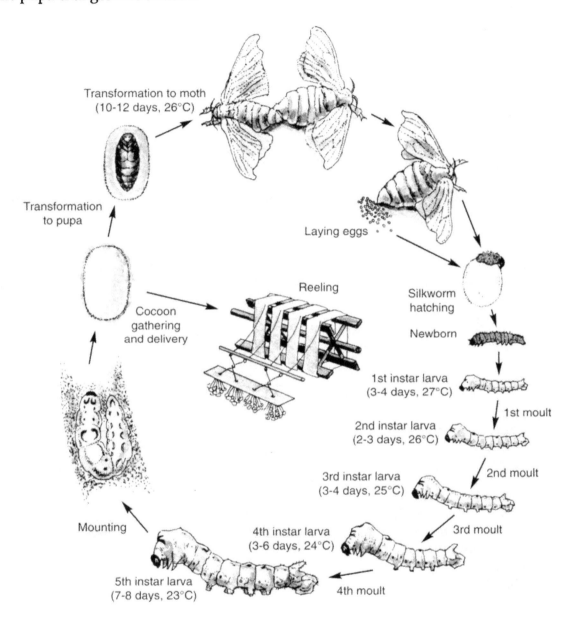

Spinning of cocoon takes 2-3 days.
Evaporation of moisture has to be controlled as it affects reeling.
Length of filament 1200 m – 1600 m and diameter 0.002 m.

Illustration 3   The life cycle of the silkworm

3

2. MORICULTURE AND SERICULTURE

Breeding silkmoths, producing disease-free eggs and rearing silkworms on a commercial scale is a complex process. Generally, hybrid eggs are reared for industrial silk production. Silkworm rearing is done by farmers who also cultivate the mulberry bushes. Developing sericulture in a new area requires the support of governmental or non-governmental agencies in creating facilities for the supply of eggs and reeling facilities for raw silk processing.

MORICULTURE (mulberry cultivation)

There are two main types of mulberry bush: white mulberry, *Morus alba*, which produces a pinkish-red fruit; and the black mulberry, *Morus nigra*, which produces a purplish-black fruit used in the production of jams or wine. The white mulberry is grown extensively for its leaf to feed silkworms, although, of 1200 varieties of white mulberry throughout the world, only a few varieties have proved to be ideal for silkworm rearing.

The white mulberry can be successfully grown in almost all climatic zones, in temperate as well as tropical areas, in rainfed or irrigated conditions and some varieties even thrive on arid, marginal land. The mulberry bush or tree, under natural conditions, has a longevity of several hundreds of years, but the cultivated mulberry, because of regular training, may only live for fifty years.

The growth of mulberry is influenced by temperature and soil conditions. In the temperate regions, mulberry buds sprout during spring when the weather begins to warm and the tree grows most vigorously during the summer months when it is hot. The growth slows down as the temperature falls and it stops towards early autumn when the tree sheds its leaves and enters dormancy. In tropical areas, mulberry has no dormant period, but shows some growth variation during dry and wet seasons in areas where there is good rainfall. In tropical regions, leaves can be harvested six times annually.

Mulberry can be propagated by methods such as grafting, layering or seeping, but the use of planting cuttings is the simplest, most economical and also most common method used by cultivators.

The quality of cocoon the silkworm produces depends on the quality of mulberry leaf on which it feeds. As the rearing of silkworms depends on the growth of mulberry, systematic planning is essential for mulberry farming and silkworm rearing. Attention should be given to timing the various operations such as ploughing, intercultivation and fertilizing.

Establishing a mulberry plantation

Mulberry belongs to the family moraceae. There are more than one thousand varieties of mulberry, including wild and cultivated forms. They are classified by the differences in the leaves, flowers and fruits. The mulberry usually bears bisexual flowers, although varieties bearing only male or female flowers or even sterile flowers are not uncommon. Because of such diversity, farmers prefer to propagate a particular mulberry variety through cuttings in order to retain the desired characteristics in the plant.

The stems of well grown plants are selected for cuttings. They are cut into 20cm lengths, each cutting having at least three buds. Pits (dimensions: 30 × 30cm and 30cm deep) are dug and filled with a mixture of farmyard manure or compost, sand and soil. The cuttings are buried and heeled into the mixed soil leaving only one bud exposed above the ground. Once rooted the cuttings are either directly planted in the field or raised as saplings in nurseries before transplanting them into the main field. Bud-grafting and root-grafting methods are adopted for exotic varieties which are not acclimatized to the local conditions.

Bud-grafting and root-grafting methods are adopted for exotic varieties which are not acclimatized to the local conditions.

Mulberry, once planted, remains productive for about 15 to 20 years so a thorough preparation of the soil with deep ploughing or digging is necessary and the ploughed land is allowed to weather. Cuttings are usually planted before the monsoon season to ensure sufficient water for their establishment in the soil. The spacing between the plants depends on the agro-climatic conditions and fertility of the soil. In favourable conditions, high plant density increases leaf yield. The normal pattern for raising mulberry bushes uses a *pit* system with pit spacing from 0.6 to 0.9m and the same distance between cuttings. In the *row* system, the distance between rows is 0.3 to 0.45m, and spacing between plants is 0.1 to 0.25m. In temperate areas, medium-sized trees are raised with 1.5 to 2.0m intervals between rows and trees.

Table 1 Spacing for mulberry cultivation

	Spacing between rows (m)	Spacing between plants (m)
Pit system	0.6 to 0.9	0.6 to 0.9
Row system	0.3 to 0.45	0.1 to 0.25

Farm maintenance

Proper maintenance of the mulberry farm is very important, as for any agricultural crop, in maintaining productivity. Intercultivation, practised regularly, loosens the soil, allows water to seep deeper, and facilitates better aeration. Timely weeding prevents soil nutrients being used up by unwanted weeds. Mulching the soil with cut grass, straw and pruned mulberry twig helps in controlling weeds, in retaining soil moisture and in protecting plants from winter injury. Good drainage is necessary to avoid waterlogging which is injurious to the plants and to the worms which feed on this leaf. Although mulberry is hardy and drought-resistant, irrigating at regular intervals in the dry season increases its yield considerably.

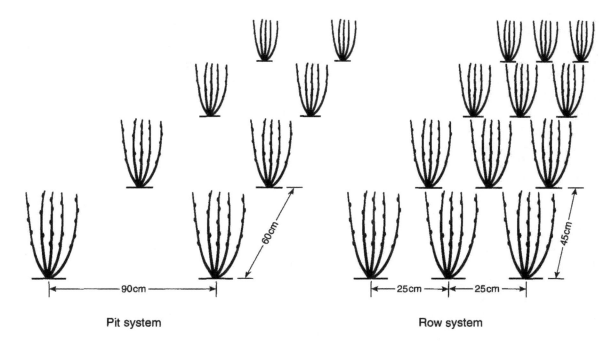

Pit system   Row system

Illustration 4 Mulberry bush spacing

After every harvest of leaf, the soil is replenished of its lost nutrients with chemical fertilizers composed of the three major elements, nitrogen, phosphorus and potassium. Nitrogen is vital and important for increasing leaf yield.

Farmyard manure and compost are regularly applied to maintain the humus content in the soil, to keep the soil structure, to improve its water-retention capacity and to reintroduce the micronutrients it has lost. Farmyard manure can also be substituted to a certain extent by green manure crops such as sun hemp, soya bean and other leguminous plants.

Mulberry bushes need regular trimming, to remove the unwanted branches, to improve the leaf yield and to improve the nutritional value of the harvested leaf. When the number of shoots is low, yield is reduced. In order to train plants into bushes and to maintain a constant number of shoots of about 80 000 per hectare, shoots are cut at the base of each plant at least once a year.

Mulberry is more resistant to the diseases and pests which cause damage to other agricultural crops. Nevertheless, mulberry maybe affected by viral, fungal and bacterial diseases, and also by physiological disturbances. All parts of the plant, including the root, stem and leaves, are prone to infections which sometimes may culminate in the plant dying. Most of the diseases can be checked by good management but some severe symptoms need to be controlled by spraying. It is recommended that advice on the types of spray to use should be obtained from local agricultural experts.

Harvesting

Mulberry leaf is harvested directly from the trees or bushes in the plantations and it is stored carefully to keep it fresh for feeding the silkworms. There are three methods used in harvesting the leaves:

- ❏ individual leaf picking
- ❏ branch cutting
- ❏ whole shoot harvesting

In the first method individual leaves are plucked from the plant, while in the latter two methods, the branches and entire shoots are cut and fed to the worms. These methods help in training the bushes. As the quality of leaves is affected by withering, it is important to harvest leaves during the cool hours and to preserve them with care.

As the nutritional requirements of young silkworms are different from those of older ones, tender leaves near the tip of the branch are picked for the young worms and more mature leaves are fed to the older worms. After each harvest, the plant is allowed to rest for a period of 35 to 45 days without picking leaf. Cultivation operations like digging, fertilizer application and weeding are repeated after each harvest. Farmyard manure is applied at least twice a year.

Table 2 Mulberry leaf harvest schedule in tropical regions over a period of 13 weeks

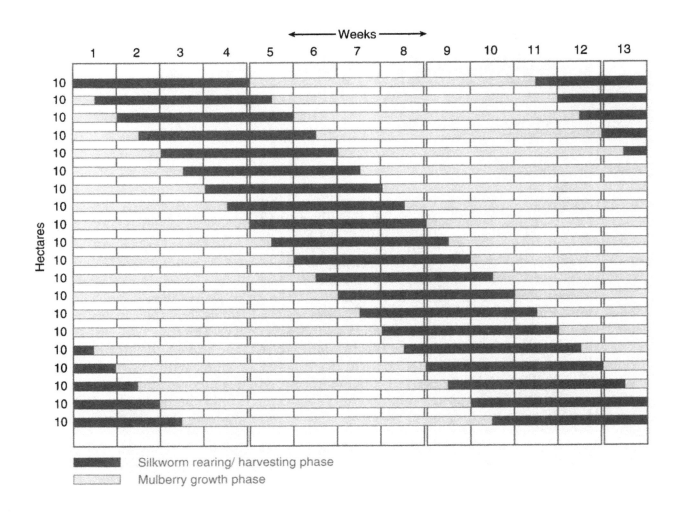

Table 3 Mulberry leaf harvest schedule in temperate regions over a period of 12 months

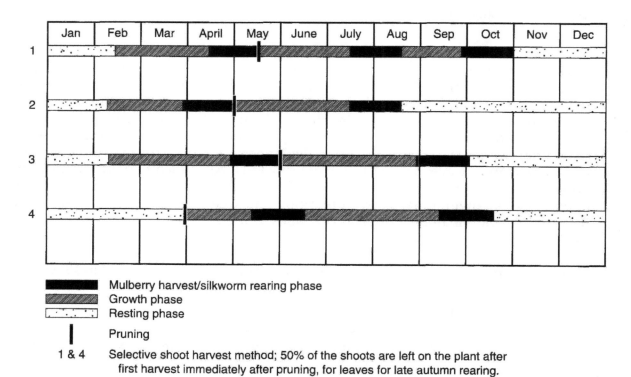

	Jan	Feb	Mar	April	May	June	July	Aug	Sep	Oct	Nov	Dec

Rows labelled 1, 2, 3, 4.

■ Mulberry harvest/silkworm rearing phase
▨ Growth phase
▒ Resting phase
▎ Pruning

1 & 4   Selective shoot harvest method; 50% of the shoots are left on the plant after first harvest immediately after pruning, for leaves for late autumn rearing.

SERICULTURE

Types of silkmoth

The silkmoth, belonging to the order *Lepidoptera*, is of the same family as moths and butterflies. As a result of controlled rearing and breeding, silkworms are now completely domesticated. Domestication has caused a degradation of senses and instincts in these insects. There are no indigenous, wild *Bombyx mori* in the world, and left to itself, the domesticated silkworm would not survive.

Rearing and breeding of silkworms in captivity has given rise to many varieties which can be classified according to their life cycle. Some silkworm varieties produce eggs only once a year, some twice and others several times, and they are called *univoltine*, *bivoltine* and *polyvoltine* or *multivoltine* respectively. The first two varieties are found in temperate regions while the latter type is found in the tropics. In temperate regions, hatching of silkworm eggs is synchronous with the mulberry growth cycle. Univoltine eggs hatch during spring when mulberry starts budding after a prolonged winter. After summer, eggs lie dormant until the next spring. The eggs of the bivoltine variety hatch in spring and become moths in the early summer, laying eggs which are similar to polyvoltine eggs. These eggs do not rest dormant, but hatch within ten days of laying and the second batch of larvae grow into adults during early autumn. The eggs laid by this second batch are dormant until the following spring, resulting in two generations every year. The eggs of polyvoltine varieties will not become dormant at all, but hatch after ten days of laying.

8

Silkworm classification

Classification of varieties is also based on the number of moults a silkworm undergoes during its larval stage. The silkworm body is covered with a non-living substance known as *chitin*. During the growth process of a silkworm, this coating can expand only to a limited extent and has to be shed before the worm grows bigger. The casting of old skin after a new one is formed beneath is known as *ecdysis* or *moulting*. Normally a larva moults four times, but since the process is controlled by hormones, occasional deviations occur, resulting in three moults or five moults. The silkworms are correspondingly named as *tri-*, *tetra-* and *penta*-moulters. Neither tri-nor penta-moulters are desired by sericulturists as rearing them has no economic benefits.

Silkworms are often categorized as Japanese, Chinese, European and so on, depending on their country of origin. Their place of origin can be traced back through certain characteristics; for example, Japanese races have uni- and bivoltine varieties producing medium-size larvae and peanut-shape cocoons; Chinese races produce small larvae and shortened oval cocoons; European varieties produce the largest larvae and oval cocoons; and tropical races produce spindle-shape cocoons.

Silkworms are also categorized by the pattern of their cross-breeding. For commercial purposes, most of the silk-producing nations only make use of hybrids, which are obtained from crossing two original parental strains or pure races. Triple hybrids are crosses of three races in which one parent is a hybrid and the other a pure race. Double hybrids are those in which both parents are hybrids of different varieties. Table 4 shows examples of different hybrids from A, B, C and D varieties.

Silkworm varieties can be classified by their cocoon colour and larval markings. Cocoon colours range from white through to golden-yellow, orange, pink and green. The markings on the larvae are identified as normal markings, plain, zebra markings, lateral stripes etc. Generally, there are two pairs of markings of half-moon shaped spots, one pair each on the fifth and eighth segments. There are, in addition, two semicircular marks on the thorax referred to as *normal marking*.

Table 4 Examples of silkworm hybrids

A × B	F1 hybrid
A × (B × C)	Triple hybrid
(A × B) × (A × B)	F2 hybrid
(A × B) × (C × D)	Double hybrid

SILKWORM EGG PRODUCTION

The general practice followed by farmers who cultivate mulberry is to rear silkworms which, in the form of cocoons, are later sold. The silkworm rearers are supplied with eggs by a different set of sericulturists. Silkworm eggs are also known as *seed*. The egg producers procure parental seed cocoons from the seed rearers and process them in their egg-production centres.

Silkworm eggs

Each silkworm egg is about 1mm wide and 1.3 to 1.4mm in length, weighing about 0.5mg. The weight and size of the egg varies depending on its parent varieties and the environment in which it is grown.

Eggs for commercial rearing are available either on egg cards or as loose eggs. The traditional method of transporting eggs to market in China, Japan and Brazil which is now being adopted in India is by box. A box, measuring about $200 \times 100 \times 20$mm and made of a thin wooden frame covered with cotton gauze on both sides, usually contains about 20 000 eggs weighing 11g (Illustration 5). In tropical countries such as India, egg cards are still used extensively (Illustration 6). Commercial eggs are usually hybrids of two unrelated races of silkworms. Silkworm eggs are produced in a seed production centre known as a *grainage*, derived from the word *grain* meaning seed. See Chapter 5 for technical details on planning a grainage facility.

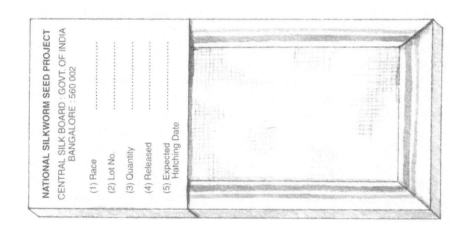

Illustration 5 Silkworm egg box (containing 20 000 eggs)

Egg selection

Original strains or pure races of silkworms are produced by farmers who breed selected seed cocoons. Special care is taken by these farmers to rear the parent races meticulously and scientifically, in order to get healthy, strong moths for reproduction. Parent seed-cocoon crops are regularly monitored, allowing diseases to be detected in the early stages. Only after screening are healthy stocks selected for hybrid seed preparation.

Sex determination of moths

In order to produce hybrid eggs, the male and female moths of two different varieties are chosen. The sex of the individual moth is identified before it reaches adulthood and before it mates. The sex of a silk moth can be determined when it is still in its larval stage, but it is a time-consuming process and requires a high level of skill. It is usually done at the pupal stage, when the sex markings are more obvious. This means the pupae have to be taken out of the cocoon for sex separation. A few varieties have certain sex-linked characteristics such as a difference in cocoon colour or larval markings which enable sex separation without cutting open the cocoon shell.

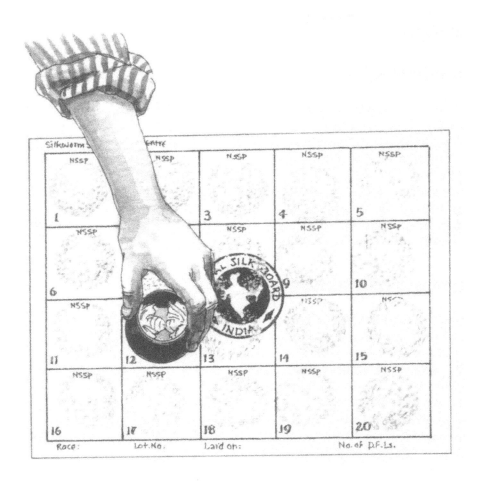

Illustration 6 Silkworm egg card (approximately 8 000 eggs)

Seed cocoons

Cocoons selected for seed preparation are stored in well-ventilated areas, where the temperature and humidity are controlled. As light has some effect on the emergence of moths, cocoons are exposed to regular cycles of light and dark. For producing hybrid seeds, it is necessary to ensure emergence of both varieties on the same day. The timing of emergences can be manipulated by one or two days by subjecting the cocoons or moths to cold storage.

Mating and egg laying

The moth normally emerges from the cocoon shell during the early hours of the day. Mating takes place between selected male and female moths under controlled environmental conditions. The male moth is attracted to the female by its *pheromone*, a scent which is present in all insects, but which is specific to each species. Female moths are separated after a few hours of copulation and are kept in the dark for egg laying. They are either confined individually in dark rings for preparing cards or collectively in an enclosure for producing loose eggs. The female moth lays between 400 and 600 eggs overnight.

Moth examination

The female moth completes egg laying within 16 hours. After it has finished laying, the moth is killed in order to detect infection or disease. The body fluid of the crushed moth is examined for the occurrence of the disease caused by a *protozoan*, which is generally transmitted through ovarian infections. The affected female moth transmits the *pathogen* into the eggs and hence to the next generation. If the eggs laid by an infected female are identified, they are destroyed. The disease can be detected earlier in the pupal stage by examining the midgut of the *alimentary canal* of a few sample pupae. This procedure is important as it ensures the production of disease-free layings, known as DFLs. About 50 DFLs make up one box of eggs.

Incubation and cold storage

The polyvoltine eggs normally hatch after ten days. For uniform development of the embryo and uniform hatching, the eggs are incubated at a temperature of 25°C and at about 75 per cent relative humidity. They are exposed to at least 12 hours of light daily which facilitates uniform hatching. In temperate regions, eggs laid in autumn undergo dormancy. They experience slow cooling as winter approaches and after the hibernation period they slowly respond to the coming of spring. These conditions are stimulated artificially in cold storage so that eggs are made to hatch after 4, 6, 10 and 13 months as required. Dormant eggs can be made to hatch like polyvoltine eggs within ten days of laying by hot or cold acid treatment.

SILKWORM REARING

The silkworm rearer aims for a good yield of the best quality cocoons which will have the best market value. All techniques and practices are aimed at obtaining this result with minimum labour and expense.

The rearing house

Unlike the rearing of *eri*, *tussah* and *muga* silkworms which is done outdoors, sometimes in the forest, mulberry silkworms are reared indoors. The majority of sericulturalists in India and Thailand carry out rearing within their houses. Separate buildings constructed for rearing silkworms have their advantages but require capital to build.

Since environmental conditions have a direct influence on the yield of a cocoon crop, the construction of a rearing house is critical. In tropical areas, where extra heating is not necessary, a simple structure with a roof to protect against sun and rain is sufficient. The rearing house should also provide protection against the *uzifly*, lizards and rats, and should facilitate cleaning and disinfection. The same building is often used for rearing young and grown silkworms, for cocoon spinning and for leaf storage. Within the rearing building, silkworms are reared on trays or shelves. The place where silkworms eat and sleep is called a *rearing bed*.

Growth of the larva

At the time of hatching, a silkworm larva is about 2.5 to 3mm long and is covered with black hair. After 25 to 30 days feeding, the larva is ready to make its cocoon. During the larval stage, the silkworm casts off its skin four times as it increases in size. While preparing to cast its skin, the larva stays still for more than 24 hours without eating mulberry. Special care is required in handling moulting and newly moulted larvae. Between the moulting stages, the larva passes through the eating periods which are identified as 1st instar, 2nd instar, 3rd instar and 4th instar until the final 5th instar, from which the silkworm enters into a cocoon and subsequently into the pupal stage. The first three instar larvae are usually referred to as young age larvae or *chawki* worms and the 4th and 5th instar larvae are described as grown silkworms. The nature of the *chawki* worms differs from that of the grown worms, and it is therefore necessary to use different rearing techniques for each instar.

The growth of silkworms is greatly affected by the environmental conditions. Normally, high temperature stimulates the growth rate and low temperature retards it. However, too high or too low temperatures are not tolerated by the silkworms and the health of the larvae may be affected. Optimum conditions for the growth of young silkworms are 26 to 28°C and 90 per cent humidity, and grown silkworms require 23 to 25°C and 70 to 80 per cent humidity.

Under normal circumstances, the silkworm increases by 10 000 times its body weight, from the time of its hatching to the spinning stage.

Table 5 Rate of increase in body weight of silkworm during growth

Growth stage	*Increase in weight* (hatched weight = 1)
Immediately after hatching	1
2nd instar after moult	10 to 15
3rd instar after moult	75 to 100
4th instar after moult	350 to 500
5th instar after moult	1800 to 2200
Full size	8000 to 10000

Planning a rearing house

In order to get a good yield of cocoons, it is necessary to plan carefully the entire rearing process beforehand. In tropical countries a box of silkworm eggs (to produce about 20 000 larvae) requires about 500 to 550kg of mulberry leaves. An estimate of the mulberry available in a garden is the most essential information required for planning the rearing operation. As the whole procedure is labour intensive, the organization of labour and delegation of responsibilities needs to be carefully arranged. In India, for instance, at least 40 to 50 labour hours is required to produce 10kg of cocoons.

The rearing space required mainly depends on the leaf availability, the number of larvae to be reared at a time, and the method of rearing. The three methods are:

❑   shelf rearing
❑   platform rearing
❑   floor rearing

Most of the sericulturists in tropical areas adopt the shelf rearing method which economizes on space.

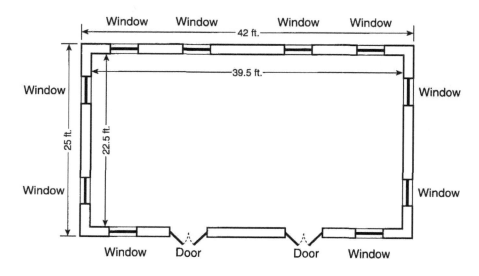

Illustration 7 Plan of a rearing house

The ideal rearing house (Illustration 7) is one which is rodent proof, well-ventilated and consists of a room for chawki rearing, one for storing mulberry leaf, another to be used as a mounting room, a bigger room for rearing grown worms and one for storing appliances. In tropical countries the rearing house should be constructed with an east-west orientation to avoid direct sunlight. The house should be oriented in a north-south direction in temperate regions.

Low-cost rearing equipment can be manufactured locally from a variety of available materials. Often the production of all the ancillary equipment can develop into small cottage industries, each with its own employment and sustainable income potential. Illustrations 8 to 14 show some of the basic equipment required for shelf rearing. It should be available and ready to be installed before rearing begins.

Wooden or bamboo, 2.5 m high × 1.5 m long × 1.0 m wide, 10 shelves, 20 cm space between shelves

Illustration 8 Rearing stand made of wood or bamboo

Round, woven bamboo, 1.2 m diameter × 10 cm deep or rectangular, wood, 0.75 × 1.0 m and about 10 cm deep

Illustration 9 Rearing tray

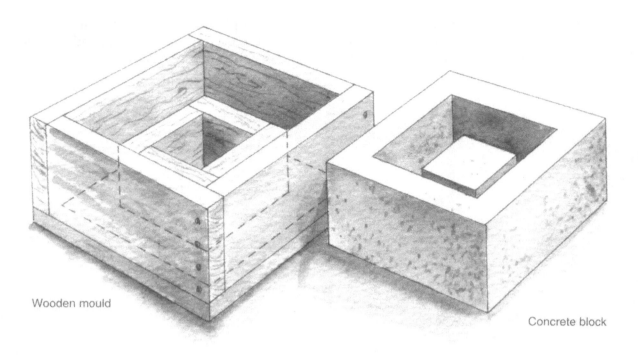

Wooden mould

Concrete block

Cast concrete blocks, 20 cm square with 7.5 cm high wall in centre, from wooden mould

Illustration 10 Ant well

Wood, 0.9 m long × 0.7 m wide × 7.5 cm deep; bottom of 2 cm × 1 cm wooden strips with 1.5 cm gaps

Illustration 11 Chawki box

Wood, 0.9 × 0.9 m × 5 cm thick and knife about 50 cm long with broad blade

Illustration 12 Chopping board and knife

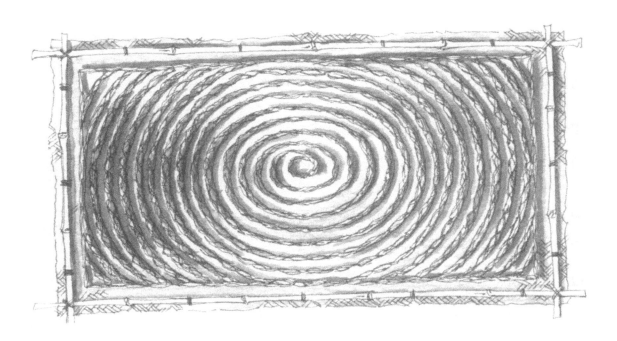

Indian bamboo chandrike, 1.8 × 1.2 m, supported on bamboo stick frame, with 5 to 6 cm high woven bamboo walls fixed in spiral with 5 to 6 cm space between.

Alternative mountages: *centipede rope, plastic bottle brush, straw frames, cardboard egg crates or twigs*

Illustration 13 Mountages

Wood, 0.9 m high

Illustration 14 Folding feeding stand

Ancillary equipment for rearing

The following equipment is also required:

❑ cleaning net made of cotton or nylon with mesh to suit size of silkworm at different instars

❑ 50 × 50 mm, 0.5 m long foam rubber or plastic strips

❑ feathers for brushing

❑ chopsticks for moving leaf at *chawki* rearing stage

❑ paraffin wax paper to line bottoms of rearing trays

❑ foot disinfection trays

❑ jute sacking or gunny cloth

❑ washbasin

Silkworm brushing and *chawki* rearing

Silkworm eggs are produced at the grainage and are available to the rearer either as loose eggs in a box or stuck on to egg cards with the natural gum produced by the female moth at the time of laying. Brushing is the process by which silkworms are transferred to their rearing bed and given their first feed. The method of brushing differs according to how they were transported.

For every box of 20 000 eggs about 100g of fresh, tender leaf is chopped to a size of 1×1cm for feeding the newly hatched silkworms. The cotton gauze on one side of the egg box is cut on two adjacent sides, forming a corner which is opened to expose the newly hatched larvae within. The box is placed in a flat-bottomed tray and a net with a mesh size of 2 to 3mm is spread over the box on to which the freshly chopped leaf is sprinkled. The box is then covered with a sheet of paraffin wax coated paper. After about six to eight hours, the larvae will have moved on to the top of the net where cut leaf has been placed. The net is transferred carefully on to a rearing tray, on the bottom of which a paraffin paper is spread to form the rearing bed. More chopped leaf of the same size is sprinkled and once again the bed is covered with a sheet of paraffin wax paper.

Where egg cards are used, chopped leaf is sprinkled directly on to the egg cards. After a few hours, the leaf bits along with the larvae clinging to them are gently brushed on to a sheet of paraffin wax paper in the bottom of a rearing tray.

The *chawki* silkworms need careful handling, as the success of the cocoon crop mainly depends on the way the young worms are treated. Factors influencing the health of the worms are leaf quality, environmental conditions and general hygiene. It is important to rear younger silkworms in a very clean environment, and to feed them with adequate quantities of fresh, tender and nutritious leaf.

The leaf is fed to the larvae as often as three to five times daily. Increased humidity in the rearing bed keeps leaf fresh for longer. The ideal conditions for the healthy growth of *chawki* silkworms is to maintain a temperature of 26 to 28°C with 90 per cent humidity. This temperature and humidity is reduced, after the third instar, to a minimum of 25°C and 75 per cent humidity. Younger worms are susceptible to high concentrations of carbon dioxide and, as the rearing of *chawki* worms is often conducted in closed, warm and wet conditions to prevent mulberry withering, it is necessary to let fresh air in at least three times daily to prevent the build-up of this gas.

The growth rate of a worm is at its highest during *chawki* stages. Although feeding is important, overfeeding causes piling up of uneaten leaf, which is bad for the worm's health. Only 25 per cent of the leaf required by the worm is eaten during its first three instars. The size of the chopped leaf is progressively increased according to the growth in size of the larva.

Mulberry leaf for the young silkworm should be tender, succulent and nutritious. The right leaves are identified by their location on the shoot. To identify them, the upper part of the shoot is held lightly and the hand is moved upwards (see Illustration 15). The leaf which stands out to the last is identified as the largest glossy leaf. The glossy leaf can also be identified by holding the terminal bud lightly between two fingers and bending it downwards. The leaf that stands out at right angles to the tip is the largest glossy leaf. The third and fourth leaves below the largest glossy leaf are fit for feeding the first instar larvae; for the second instar larvae, the fourth and fifth leaves should be used; and for the third instar larvae, the fifth and sixth leaves are the best.

Table 6 Leaf requirements for one box of eggs

Growth stage	Quantity of leaf	Size of chopped leaf
1st instar	350g	0.5 to 1.0cm square
2nd instar	1500g	1.5 to 2.5cm square
3rd instar	7000g	3 to 4cm square

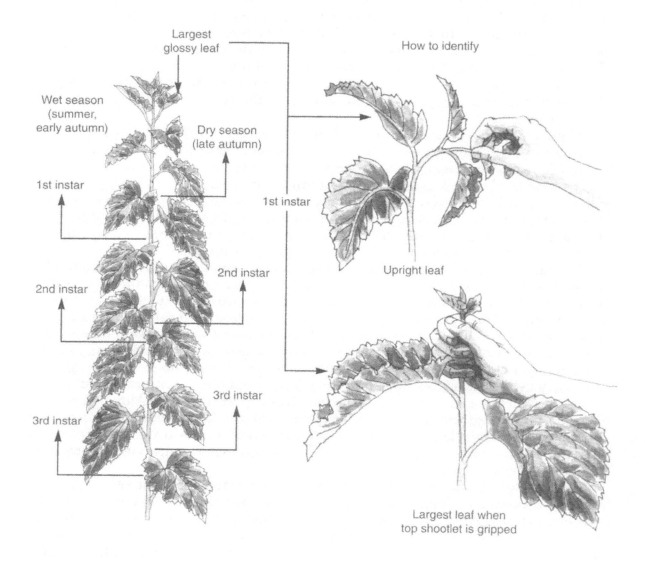

Illustration 15 Leaf selection for each stage of silkworm growth

The silkworm slows down its eating pace on the third day and its body appears shiny, which indicates that the worm is ready to cast off its skin. To give it a clean place for moulting, the rearing bed is cleaned of old leaf, chemicals and other debris. When nearly 75 per cent of the worms in the bed have stopped eating, feeding is stopped. Slaked lime is sprinkled on the bed to dry the remaining leaf. The paraffin wax paper covering the beds is removed, and the trays are kept in a well-ventilated room. Feeding is resumed after about 20 to 24 hours, when the majority of the silkworms in the bed have moulted.

Bed cleaning is done prior to moulting, immediately after moulting and at least once daily during feeding stages. For cleaning, a net sprinkled with fresh mulberry leaf is spread on the bed. The silkworms come up through the net to eat the fresh leaf and as they feed the net is transferred to another tray to make a new bed.

As the worms grow in size, the bed space becomes crowded and the worms tend to compete for food. The bed should be enlarged to accommodate their growth. Increasing the bed space and cleaning are usually done simultaneously.

Table 7 Space required for different instars of larvae from one box of eggs

Age	Space needed at the beginning of instar	Space needed at the end of instar
1st instar	$0.2\,m^2$	$0.5\,m^2$
2nd instar	$0.5\,m^2$	$1.5\,m^2$
3rd instar	$1.5\,m^2$	$3.0\,m^2$

Rearing 4th and 5th instar silkworms

The 4th and 5th instar worms differ from those of earlier stages in many respects. They are susceptible to high temperature, high humidity, and poor ventilation but can, to a certain extent, tolerate poor quality mulberry leaf. Good ventilation is necessary to displace the bad air breathed by thousands of fast-growing worms in the rearing room. Conversion of leaf protein into silk within the silk glands occurs in these instars, so the leaf fed to the silkworms should have a high protein content. The grown worms are voracious eaters consuming 75 to 80 per cent of the total leaf required for their growth. Table 8 shows the feeding requirements for rearing grown worms.

A large quantity of leaf is consumed by silkworms during the 4th and 5th instars and a correspondingly large quantity of leaves must be harvested daily, which requires more labour. In order to reduce labour costs and to preserve the leaf quality for a longer duration, *shoot rearing* is preferred by many farmers. It is done indoors either on the floor or on platforms that may be on two or three tiers. Whole shoots are placed on the platforms which form the rearing bed for grown worms. This practice reduces the labour required for leaf picking and for leaf preparation for feeding.

Mounting and spinning

In about a week after the fourth moult, the silkworm loses its appetite. Its body appears translucent and a pair of silk glands are visible through the skin. The larva passes semi-solid faeces. It becomes more active and moves around impatiently, looking for a suitable space to spin its cocoon. There is a tendency for the mature silkworm to move upwards but away from light. The mature silkworms are picked up individually and put on to a mountage. If cocooning frames are used, the silkworms are allowed to mount the frames unaided.

Table 8 Requirements for rearing grown silkworms

Requirements	4th instar	5th instar
Type of leaf	Medium	Medium coarse
Leaf size	Whole leaf or shoots	Whole leaf or shoots
Temperature	25°C	24°C
Humidity	75%	70%
Cleaning	Daily	Daily
Space (1 box eggs)	3 to 9 m²	9 to 18 m²
Quantity of leaf	50 to 60 kg	370 to 420 kg

The larva exudes a viscose-like fluid through the *spinneret* situated above its mouth, stretching the silk thread across the enclosed space of the mountage in all directions, making a loose framework that provides footage for the cocoon it is going to build. For the last time, the silkworm passes urine, protruding its hind end out of the silk framework, before it actually starts making the cocoon. The silkworm draws its head back and begins to move it around in the form of a figure of eight (8) while exuding the silk fluid all the time, which forms a filament of uniform thickness. It is estimated that about 30cm of silk is exuded by the silkworm every minute. The worm crawls around within the cocoon shell building up layer after layer of silk around itself until all the silk is squeezed out of its glands. It takes about 48 hours of continuous and tireless work for the silkworm to make a cocoon consisting of a 1200 to 1500m long silk filament. In the process, the worm shrinks considerably in size. The worm then rests briefly and transforms into a chrysalis or *pupa* by casting its skin.

It is necessary to dry off the silkworm urine before it damages the neighbouring cocoons already formed in the mountage. As much as 100ml of urine is discharged by just 250 worms, which is suffecient to increase atmospheric humidity considerably when a large number of worms is reared. The mountages are kept undisturbed until the cocoons are complete and the chrysalises, or pupae, are well formed. The pupa, soon after its formation, has a very tender skin which gets damaged at the lightest touch. Dead worms and flimsy cocoons are removed. By the fourth or fifth day, when the pupal skin has sufficiently toughened, cocoons are harvested from the mountages.

Loose silk floss which forms the footage for the cocoons is removed along with the soiled, stained and double cocoons, in preparation for the market. The cocoons are loosely packed in baskets or cloth bags and are transported during the cool hours of the day.

Cocoons are sold by the rearers and are bought in the market by the reelers. This stage in the marketing of cocoons marks the end of the Agricultural Sector and the beginning of the Industrial Sector of silk production.

DISEASES AND PESTS

Cocoon yield is greatly affected by the occurrence of disease in silkworms. *Flacherie* caused by a bacterial infection and virus takes the highest toll, infecting the silkworms through the mouth. The other common diseases are *nuclear polyhedrosis* or *grasserie*, *muscardine* and *pebrine*. Infections affect the silkworms in different ways. While muscardine fungus attacks the skin, pebrine, caused by protozoa, is transmitted through the ovaries and eggs and/or through the mouth. The diseases, to a large extent, can be controlled by rearing silkworms under hygienic conditions and feeding them nutritious food, while maintaining the maintaining optimum environmental conditions. This keeps the worms healthy and allows them to develop resistance to infection. Controlling any disease once a batch of silkworm is infected is difficult and expensive, so preventive measures (like the disinfection of rearing houses and equipment) are better than combating disease once it has occurred.

Disinfection

Silkworms are susceptible to infection from *pathogens* which lurk in the rearing room and equipment. It is necessary to disinfect and kill these *pathogens* before rearing is started. Rearing equipment is initially washed in warm water mixed with bleaching powder and dried in the sun. Sun-drying, to a certain extent, disinfects the equipment. Rearing rooms are thoroughly dusted and washed before, disinfection which is done by spraying 2 per cent formalin solution inside the room and on to equipment. The solution is prepared by mixing one part of commercial grade formalin (35 per cent) with 11 parts of water. Nine litres of prepared formalin solution is required to disinfect a room of $3 \times 3 \times 3$m.

In a room that can be made airtight, fumigation is convenient, as the room and equipment are saved from wetting. Holes and crevices are sealed to prevent gas from escaping. Using a dosage of 60g per $10\,m^3$, Neo PPS or Para Formaldehyde is heated in a pan for about four or five hours. The gas emitted from this is effective in killing the disease which causes pathogens. The rearing appliances are spread in the room before fumigation to ensure thorough disinfection.

3. SILK REELING AND FABRIC PRODUCTION

Illustration 16 shows the sequence of production operations from cocoon to marketable raw silk. Also shown are the options available for cooking, brushing, reeling and baling.

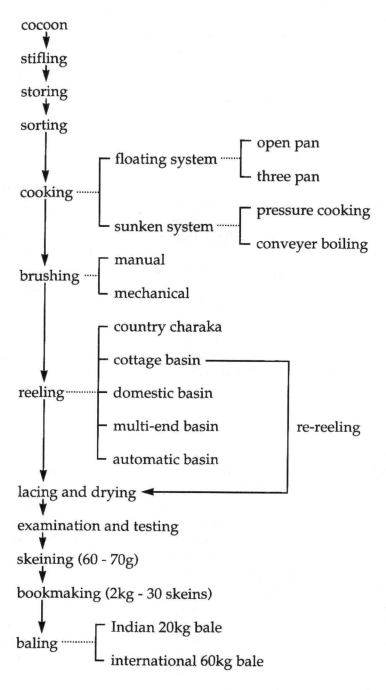

Illustration 16 Sequence of operations in silk reeling

PRE-REELING PROCESSES

The cocoons procured by the reeler are subjected to many pre-reeling processes which are necessary to increase silk productivity and to improve its quality, assuring the reeler of higher returns. Cocoon characteristics influence silk quality to a great extent. A reeler's success is dependent mainly on an ability to assess the quality of cocoons, and to price the lot purchased. The reelers assess the quality of cocoons by studying certain characteristics in the batch.

Most of the weight of a cocoon is taken by the pupa which has no commercial value in reeling. The cocoon is covered by a loose, entangled filament called floss, under which the cocoon shell is well formed. The shape of the cocoon and its colour is peculiar to the race. The cocoon surface has wrinkles called grains which are more pronounced in some varieties. Fine-grained cocoons are more compact and yield better silk. The weight of the cocoon varies with different breeds and within the same race in different seasons. A cocoon progressively loses weight from the day it is spun, because the stored food is slowly used during the respiration of the chrysalis inside.

There are usually many defective cocoons in a batch. Bad cocoons affect reeling efficiency and the quality of the silk. Black cocoons (soiled by dead pupae), dead cocoons, and cocoons spotted or stained by urine of other worms are separated from the good ones. If these bad cocoons can be reeled at all they are normally reeled without any pre-reeling process and the raw silk produced from them is sold more cheaply.

Cocoon drying and stifling

The object of cocoon drying is to prevent the emergence of moths, and to remove some of the moisture contained in the cocoon shells and pupae, so that they can be stored for some time. Various methods are adopted to kill the pupae.

Steam stifling

Steaming is the simplest method of stifling (Illustration 17). Bamboo baskets are loosely filled with live cocoons and the baskets are placed on a low platform inside a metal drum. The drum is partially filled with water and is heated from below. It is covered with a lid. Hot steam circulating in the basket kills the pupae within a few minutes. The cocoons are removed from the baskets and are spread on a mat to allow them to dry before being stored. In large reeling units, steam is supplied to a chamber in which the cocoons are arranged in trays and kept on racks.

Steam stifling is more appropriate in tropical countries where cocoons are produced all the year round. It is not appropriate for temperate areas where cocoon production is restricted to a particular season, as the reeler has to store cocoons for longer periods. Steam stifling produces cocoons which are not completely dry, and because of the presence of moisture, the cocoons develop moulds when they are stored for long periods.

Hot air drying

Hot air drying is used by some reelers who prefer to keep their cocoons dry. In addition to killing the chrysalises, moisture does not penetrate the cocoons.

A simple hot air drier has a chamber containing shelves on which the cocoons are arranged. A fan at one end maintains the air current which is heated by a stove and driven into the chamber. Hot air blown over the cocoons kills the pupae and at the same time dries them. A total of 714 calories of heat is required to dry one kilogram of cocoons.

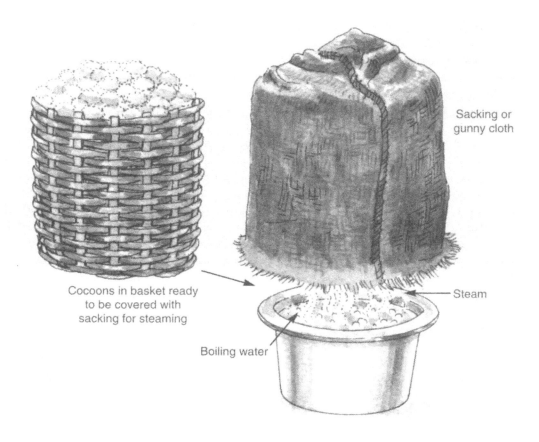

Cocoons in basket ready
to be covered with
sacking for steaming

Sacking or
gunny cloth

Steam

Boiling water

Illustration 17 Steam stifling

Conveyor belt drier

The conveyor belt method adopts the same principle as that of the hot air drier. With the hot air drier the temperature to which the cocoons are exposed is not regulated. The conveyor drier is a more sophisticated piece of equipment with both temperature and the duration of drying automatically regulated to give the best result. The drier is a rectangular chamber containing four tiers of wire mesh conveyors on which the cocoons are placed. Hot air is supplied by a fan from a radiator, which provides different temperatures at different parts of the chamber. Hot air rising through the layers of cocoons takes away the moisture. The temperature at each tier of the conveyor and the duration for which cocoons are subjected to a particular temperature can be decided and regulated by various controls.

Completely dried cocoons come out of an outlet at the bottom. Dried cocoons are stored in a well-ventilated place to avoid mould and insect attack. All defective cocoons are removed, which otherwise would attract *dermestid beetles* and other pests.

Cocoon sorting

Only at the time of reeling are cocoons sorted. Bad cocoons hamper the reeling process and affect the quality of silk yarn produced. Small quantities are sorted at a time. Thin-shelled, double and perforated cocoons must be identified and removed. Generally, floss is removed by hand, or with the help of a simple device known as a deflossing machine. The machine has a rubber sheet with firm ridges, rolled between two rollers. The cocoons are fed through a conveyor, the floss gets caught between the rollers and is teased out, leaving cleaned cocoons to pass into a container.

Cocoon cooking

Prior to reeling, the cocoon is cooked in water to dissolve the soluble protein serecin which sticks the cocoon filaments together in the shell. The cocoon softens and swells. The filament can be easily taken out and wound on a reel.

The cooking bath is a copper, brass or aluminium vessel of 30cm diameter. Water in the bath is heated either by burning fuel underneath it, or by steam supplied through pipes from a boiler.

Many methods of cooking are adopted by the reeling mills which are also known as *filatures*. The simplest method of cooking is by immersing cocoons in boiling water. After boiling them sufficiently, the cocoons are transferred to the reeling basin with a perforated ladle. To get the reelable end of the filament, the cocoon surface is brushed gently with a bamboo stick or a straw bunch when the cocoon is still in the cooking pan.

A slightly improved method is to immerse cocoons, loosely packed in a metal cage, for a few minutes in boiling water, and then in cold water. Air is expelled from the cocoons during the hot water treatment which makes them absorb water inside the cold water bath and they become heavier and are easier to reel. Cocoons with very hard shells require pressurized cooking so that the shell is cooked uniformly. Some equipment is specially designed to immerse the cocoons in water at different regulated temperatures for different durations, by passing them in cages into water baths. After cooking, the cocoons are subjected to mechanized brushing, in which a circular brush with straw bristles arranged on the lower surface is immersed in the water bath, where the pre-cooked cocoons are collected. The brush is rotated, changing direction frequently, to brush off any loose filament and bring out the reelable end of the required filament.

REELING

Reeling is the process by which the filament forming the cocoon shell is unravelled. As each filament is very thin, delicate and by itself not suitable for weaving, filaments of several cocoons are reeled together to get a firm and strong yarn. The number of filaments determines the thickness or the denier of the reeled silk yarn. Reeling involves many processes and uses a variety of traditional and new technologies. For reeling devices, see *Appendix 1*.

A typical reeling device consists of three main parts, a *reel* or *swift* on which silk is wound, a *distributor* system which distributes the silk in a regular pattern on the reel and a *croissure* system which provides the necessary tension on the yarn to make the silk a strong and uniform thread.

The brushed cocoons with cleaned filaments ready for reeling are transferred to the reeling basin with a perforated ladle. A specific number of cocoons required to make the raw silk yarn are joined together and the filaments are passed through a hole known as a *threader* or *thread guide*. The yarn is wound on the reel which turns on its horizontal axis while it is drawing silk from the cocoons.

Once reeling begins, a constant number of cocoons in the group is maintained to regulate uniformity in the thickness of the yarn reeled. More cocoons are added when the number in the group is reduced as silk is drawn from them. Water in the basin is regularly changed to remove any stain or sediment and to keep the silk reeled fresh and clean.

Above the basin a frame is fitted with threaders through which the filaments taken from the cocoons are passed. The frame is fitted with *croissure* pulleys above which the reel is placed. Yarn from the guide is passed over the croissure pulleys and on to the reeling swifts. The thread-guide eye is made of a convex porcelain disc or button with a hole pierced in the centre, through which the thread passes. Modern machines are equipped with size detectors for maintaining uniformity in yarn thickness. Normally a reeling basin has provision to reel six yarns at a time although this may vary depending on the size of the basin. Each unit with a full complement of thread guide, *croissure* and traverse guide is referred to as an *end*.

Croissure

Croissure is the crossing and twining of several silk filaments as they pass between the thread guide to the reeling swift. One of the most important reasons for doing this is that it allows aglutination of the filaments of several cocoons to form a compact silk yarn. It also squeezes out water from the yarn as it is being reeled. To a certain extent, it controls quality, as weak and defective portions of filament break under its tension. There are many methods of developing a croissure. The most common croissures are the *chambon* (French) or the *tavelette* (Italian). See Illustration 18.

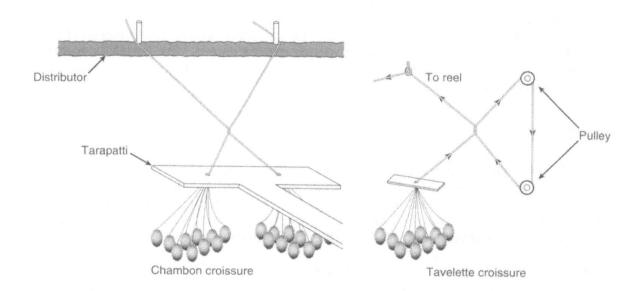

Illustration 18 The chambon croissure and tavelette croissure

The chambon croissure differs from the usual system in that it requires at least two separate threads. The two threads are intertwined for a number of turns and then passed on to separate reels, the thread from the right passing to the left reel and the left thread passing to the right reel. It produces a less advantageous light tension when producing high quality raw silk.

In the tavelette croissure, the thread runs around three small pulleys. The yarn is passed over the upper pulley and then over the lower one. During its course from the lower to the third pulley on top, the yarn is entwined on to itself. The thread from the third pulley is then passed on to the reeling swift.

Re-reeling

To facilitate packing, raw silk has to be wound on to a standard reel to make hanks of a certain length, width and weight. The process of re-reeling releases the strain to which the silk has been subjected at the time of reeling. A re-reeling device has a reel with a circumference of 150cm, thread guides, a distributor and a silk-drying system.

Before re-reeling, raw silk on a reeling spool is wetted. The thread is passed through guides mounted on a rod and through a distributor on to a swift. The distributor rod has a transverse motion that facilitates the formation of a flat hank with the thread arranged in specific patterns of a criss-cross or a diamond shape. This is necessary for easy unwinding of the thread during further processing. The silk is wound on the swift at a high speed of nearly 350 rpm. If moisture is not removed immediately from the yarn, it produces gum spots on the hanks so external heaters such as burning coal, steam pipes or heaters are placed near the reel.

In order to keep the diamond pattern on a reel intact and to facilitate easier withdrawal of the thread during later processing, skeins are laced at regular intervals before removing them from the swift.

The laced skeins are cleaned and examined for any defects. Gumspots, if any, are slowly teased out. Hanks are twisted into skeins by a hand-operated machine known as a skeining device. The skeins are packed into books of 2 to 2.5kg.

By-products

All the silk produced by silkworms is not reeled as raw silk. Part of the cocoon is known as floss waste. More waste, known as filature waste or jute, comes off during the brushing and reeling processes. The last layer immediately surrounding the pupa is also filature waste. Filature waste is about 25 to 30 per cent of the reeled silk. Silk waste forms the raw material for the production of spun silk, while the dead pupae are used in the extraction of oil and are used as fertilizer or animal feed.

Silk produced from double cocoons and flimsy cocoon is generally referred to as dupion silk, which is coarse and irregular. Traditional, manually operated devices with only two ends are used for reeling dupion silk. The reeling of *dupion* silk is done slowly. Cocoon casting is done with the help of a stick. Dupion reeling requires a high level of skill.

RAW SILK PROCESSING

Raw silk extracted from the cocoon passes through a series of operations, which make it fit for weaving. Most of the silk fabrics woven are loom finished. Even fabrics which require further dyeing and printing are degummed in yarn form before weaving. Degumming removes the serecin that binds the fibroin of individual filaments together in the yarn.

Throwing

Individual filaments of degummed silk become separated when they are not twisted or *thrown*. For producing a firm and closely woven fabric with good *handle*, raw silk has to be thrown in single or plied state. *Throwing* also improves the breaking strength and will increase the diameter and denier of the silk yarn.

The process known as throwing is the most important link between the production of raw silk with weaving. Without the throwing process it would be very difficult to manipulate the reeled raw silk as a textile yarn.

Throwing consists of four separate operations, each requiring special machinery:

1. **bobbin winder**
2. **uptwister**
3. **ring doubler**
4. **hank winder**

The four processes, seen in Illustration 19, show how several filaments of raw silk are combined to make a yarn. The throwing process produces a silk thread specific to a particular type of fabric. This process largely determines the appearance and feel of the finished fabric.

Bobbin winding   Twisting   Doubling   Hank winding

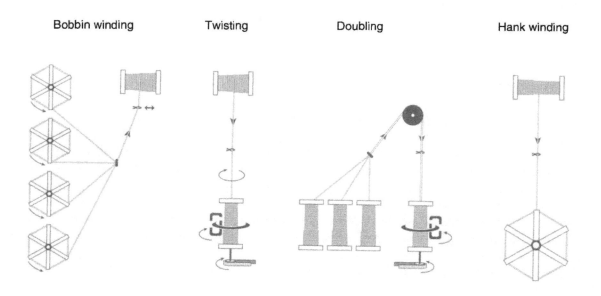

Illustration 19 The four stages of throwing

Bobbin winding

The raw silk hanks are first cleared of gum, knots and the adhering fluff. The hanks of raw silk are stretched on to *swifts* on a winding machine to produce *bobbins*. Sometimes the hank of silk is moistened to improve its strength during winding.

Twisting

There are many types of twisting machines available, the most common being the uptwister. Bobbins are mounted on spindles which are turned at high speed. There are rollers which take off the silk from the bobbins at a slower pace. The speed difference between the bobbin and the roller determines the number of twists per metre (tpm). When single yarn is twisted, it is known as 'single twist' and if the yarn is plied, it is referred to as 'double twist'. Depending on the number of yarns in the twisted thread, the thrown silk is referred to as 2 ply, 3 ply, 4 ply and so on. The twists are also described by the direction in which the yarn is twisted as in *right* and *left* twist yarns, known also as Z or S twist respectively. Illustration 20 shows how Z and S twist yarns are combined to give a balanced yarn.

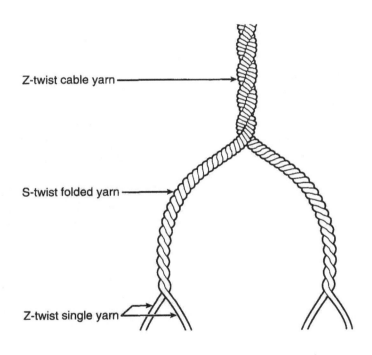

Z-twist cable yarn

S-twist folded yarn

Z-twist single yarn

Illustration 20   2 ply × 2 ply twisting or doubling

Doubling

Doubling is a process by which silk from two or more bobbins are wound together for a second time on a single twisting bobbin, which results in plying. It is done both for untwisted yarn and for threads after single yarn twisting has been done.

When the yarn has to be subjected to high twisting, the silk hanks are often soaked in soap and oil emulsion to improve their strength.

Hank winding

Once the required yarn has been manufactured it is taken off the bobbin by winding it back into a hank, so that it can be de-gummed and dyed.

Setting

The twisted yarn has a tendency to untwist and should therefore be set by steam.

Untwisted yarn or raw silk, either in single or doubled form, is used in the production of certain lightweight fabrics that require a softer feel. Low-twisted yarn is normally used as weft in the production of soft silk fabrics, which are either single or plied. Hard-twist yarn is referred to as *crêpe* and fabric made of this yarn is also refered to as *crêpe*. Often 70 to 80 tpi (turns per inch) are introduced into *crêpe* weft yarn and about 10 tpi are put into the warp yarn.

Thrown silk yarn is produced according to the type of fabric to be woven. It has its own identity according to its use: *tram*, *organzine*, *compenzine* and *grenadine*. See *Appendix 2* for a further description of these fabrics.

Weaving

It is beyond the scope of this handbook to describe the entire weaving process. The basic principles of weaving are the same, no matter whether it is cotton, wool or man-made fibres which are being woven.

Weaving is covered in more detail in the Intermediate Technology textiles handbook entitled *Fabric Manufacture*. For silks a few additional facts need to be considered. The main feature which distinguishes a handloom suitable for silk weaving from that for weaving other fibres is the required minimum distance of one metre from the back of the shafts to the back beam on to which the silk warp is wound.

Shuttles used for weaving silk are slightly different from those used for cotton or wool weaving. To prevent the silk yarn from running out too quickly, a small piece of fine fur, to act as a *drag*, should line the inside of the shuttle where the pirn is placed.

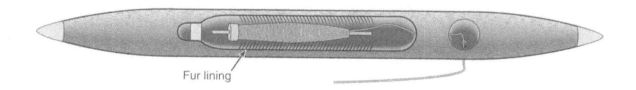

Fur lining

Illustration 21 Shuttle for silk weaving showing fur lining

DYEING

Serecin present in the silk makes it hard and stiff and its lustre and whiteness is not revealed. Removing serecin allows silk to take dyes better and to get the desired lustre, smoothness and softness in the silk fabrics.

Silk is first degummed by using soap and then washed in dilute sodium bicarbonate solution. To get a better lustre it is also treated with formic acid. Silk has affinity to various dyes such as direct dyes, acid dyes, metal-complex dyes, milling dyes and, to a certain extent, to reactive dyes. Normally the direct dyes are used to produce brown, black and bottle-green shades. Certain dark shades such as violet, blue and green are produced using acid dyes. The most popular shade of turquoise blue is obtained by using acid dyes. Certain treatments on silk improve the fastness of direct dyes against sunlight and washing. Acid-milling and metal-complex dyes are extensively used for pale and pastel shades. Basic dyes are good when brilliant shades are required although they do not provide the required colour-fastness.

Reactive dyes are gaining popularity because of their fastness, although they do not have affinity like acid dyes. Both hot and cold brand reactive dyes are available for use.

Natural silk is off-white. When a pure white fabric is desired, the yarn or cloth is bleached using hydrogen peroxide or sodium oxide with sodium silicate as a synthetic stabilizer. Regular use and washing will bring back the original off-white shade to silk.

FINISHING

In order to get certain desired effects, both dyed and printed silk fabrics are subjected to several physical and chemical treatments that are known as *finishing treatments*. When a fabric is bleached, dyed or printed, it tends to crease. Smoothing of fabrics is done by stretching them evenly and passing them over hot rollers. Stretching can also be achieved by passing them through *stenters*.

Conventional treatments on silk are *calendering, weighting, scrooping,* and *starch and glue finishing*. Many modern methods have been adopted to suit present day life styles. *Urea and phosphoric acid* finishing is used to ensure fire resistance and it also withstands all dry-cleaning operations. Some of the common finishing treatments to which silk is subjected are described below.

Sizing

Sizing is done to achieve full, firm and soft handle, giving *body* to the fabric. Starch and polyvinyl acetate emulsions are normally used as sizing agents. To maintain the effect for longer, non-ionic softeners and acrylic-based resins are also used.

Gassing and singeing

When fabric is passed over a large number of gas flames set in a row at a speed that allows only debris and loose ends to be burnt a smooth lustrous fabric is produced.

Weighting

Weighting is used to replace the loss in weight which results from degumming and can be done before or after dyeing. The fabric can be weighted with stannic-chloride or sodium carbonate without affecting the fabric lustre or feel.

4. SILK GRADING AND TESTING

Silk is produced in several countries using a wide variety of silkworms and many different technologies. Producing silk for an international market requires a common standard by which the quality of the silk can be assessed. Silk is usually judged by how clean it is and on its evenness.

The weight of silk varies under different atmospheric conditions, and is affected especially by humidity. Dealing is usually done with *conditioned weight* silk to avoid confusion during transactions.

TESTING METHODS AND CLASSIFICATION

Raw silk is generally classified by visual and mechanical tests. Usually an entire silk lot is assessed on its colour uniformity, lustre and handle before it is subjected to any other tests. Further tests are done to determine the denier of the silk and variations in its strength, elongation and winding characteristics.

Winding test

The winding test determines the productivity and quality of raw silk before weaving. The silk is wound on to a swift at a uniform speed of 110m, 140m or 165m per minute. The test will vary according to the denier of the silk. The raw silk is evaluated by the number of breaks recorded during one hour.

Size variation

The thickness of filament silk yarn is measured in *denier*. Denier is determined by the weight in grams of yarn of a fixed length of 9000m. To determine denier, the filament silk yarn is wound on to a 1.125m circumference reel. The reel, which has a counter which indicates the number of revolutions it has made, is turned at a speed of 300rpm. The silk is wound through 400 revolutions giving a 450m long hank of yarn. The weight of the hank in grams after removal from the reel is doubled to give the weight of 9000m of yarn. This weight in grams is called the denier. The weight of each hank is compared during the test to determine an acceptable denier.

Cohesion test

Silk yarn is made up of many filaments, and each filament consists of two *brins*. The silk filaments or *baves* are held together by the cohesive property of serecin. It takes a certain amount of friction to separate the yarn into individual filaments and brins. If higher friction is needed to separate them, it indicates a better quality of yarn. 'Duplan Cohesion Tester' is a device used for testing cohesion, consisting of two sets of ten hooks arranged parallel to each other at a distance. Silk yarn is wound round the hooks in a zigzag manner. The hooks hold the yarn at a constant but uniform tension. A steel blade is made to run across the threads with a uniform load and at a constant speed, so that the blade applies friction to the yarn at 20 different points simultaneously. The number of strokes required to separate filaments at any one point indicates the cohesive strength of the yarn. Repeated tests with the same batch are done to measure the average cohesion.

Tensile strength and elongation

A certain degree of elasticity in the fibre is required for easier weaving and for the durability of the fabric. Silk is more elastic than other natural fibres. The actual elongation capacity is measured using a device known as a 'serigraph'. A serigraph consists of two clamps set vertically 10cm apart. It has an attachment to record elongation and tensile strength simultaneously on a graph sheet. A constant number of threads of known denier is held by the two clamps. The clamp on the lower side is slid down to pull the threads at a speed of 15cm per minute. On reading their breaking strength, the threads snap off. Before snapping, the threads get stretched to their maximum elongation which is measured as a percentage of their total length. The load in grams per denier at the breaking point indicates the tensile strength of the yarn. The average for a number of samples represents the tensile strength/elongation value for the yarn.

Neatness, cleanness and evenness

Ultimately the texture and uniformity of a fabric is decided by neatness, cleanness and evenness of the yarn which is used in weaving. These characteristics are studied by visual methods, comparing the samples with the standards available.

To make the defects and deviations in a sample more visible, the silk yarn is wound on to a narrow, long inspection board painted black. As 'seriplane' is used for rotating the inspection board while winding the silk threads. The raw silk threads are made to lie parallel to each other in a single layer on the board to form a panel. The equipment has a counter which indicates the number of windings on each panel. A constant speed is maintained while turning and uniform tension is applied to the thread. Each board can actually take up ten panels of thread.

Each panel is about 127mm wide and 450mm long. The inspection room where the panels are compared with standard photographs is specially designed with its inner walls painted pale grey and floors and ceiling painted white. The room is illuminated with incandescent bulbs arranged in both horizontal and vertical rows. Chromium corrugated reflectors are used to give diffused lighting.

RAW SILK CLASSIFICATION

The procedure for testing and classification of raw silk is described in the International Standard Method of Raw Silk Testing and Classification. It has been officially adopted by the International Silk Association. Raw silk is divided into three categories for classification purposes, as shown in Table 9.

Table 9 Raw silk classification

Category	Denier
1	18 and below
2	19 to 33
3	33 and above

A set of grades (4A, 3A, 2A, A, B), based on test results, describes the quality of each category of raw silk.

SILK CONDITIONING

Silk is highly hygroscopic and absorbs moisture up to as much as 30 per cent of its weight. In a dry atmosphere it gives off moisture, with the result that the same quantity of silk weighs less in a drier environment. The international specified standard states that when raw silk contains the moisture equivalent of 11 per cent of its absolute dry weight, this is known as the *conditioned weight*. Since, under normal

Table 10   Tests for mulberry silk

Test	Effect on mulberry silk
Flame test	Burns slowly leaving black residue Burnt hair smell (If it is acetate rayon, there is no smell)
Diluted nitric acid at 70°C	Fibre stains yellow
Concentrated nitric acid	Disintegrates and dissolves
80% sulphuric acid	Disintegrates and dissolves
Hydrochloric acid	Dissolves
Sodium hydroxide 5% (boil)	Dissolves slowly
Millons reagent	Fibre stains red
Glacial acetic acid	No change (Acetate rayon dissolves)
Phenol 90%	No change (Nylon, rayon or cotton dissolve red to violet)
Microscopic test	Structureless Fibre rod shaped (Fibre with longitudinal striation indicates tasar silk)

circumstances, standard conditioned silk never exists, the condition has to be artificially created to ascertain its conditioned weight. A silk conditioning chamber is used for this purpose where the silk sample is dried under controlled heat. A draught of air is blown to draw away the moisture from the silk. A suspended balance weighs the silk when it is still within the chamber. A gradual reduction in the weight of silk will be noticed until the silk is absolutely dry, after which the weight remains constant. Eleven per cent of this weight is added on to obtain the conditioned weight of the silk.

TESTS FOR SILK

The simplest and most popular test for mulberry silk is the flame test. There is, however, a series of different tests which can be carried out on fabric which will also determine if it is acetate rayon. See Table 10.

5. PLANNING FOR PRODUCTION

Sericulture and silk production cover the following activities:

❑   mulberry cultivation

❑   egg production

❑   silkworm rearing and cocoon production

❑   raw silk reeling

❑   throwing

❑   weaving

To establish silk production it is first necessary to have both a reliable supply of hybrid silkworm eggs and a market demand for reeled silk.

The infrastucture for consistent and good quality silk production can be achieved by ensuring good facilities for silkworm egg production (grainage) and centralized silk-reeling units.

This chapter looks at the planning and economics of setting up these two facilities.

SETTING UP A GRAINAGE

A silkworm egg (or seed) production centre is known as the *grainage*, the word derived from *grain* or *seed*. The success of an industry is dependent on the quality of eggs made available to farmers. In a new sericulture area, where silkworm eggs are not available locally, farmers must depend on an outside source. Although procurement from outside is inevitable, at least initially, it poses problems in the long term. For the sustainability of the industry, self-sufficiency in hybrid seed production is essential. Responsibility for seed production can be taken up by individuals, non-governmental organizations or government agencies.

The primary objective of a grainage is to produce eggs which are entirely disease free. This is achieved by producing eggs from healthy and robust parents. Before planning a grainage, one has to assess the need for its establishment. Mulberry cultivated on a minimum area of 500 acres of land can support a small viable grainage unit producing 50 000 disease-free layings (DFLs) a month. Before starting on this venture, the following aspects need to be considered.

Location

The region should be basically a sericulture area with a large number of farmers practising silkworm rearing so that the demand for seeds exists, preferably throughout the year. The climate should be conducive to egg production. The ideal conditions of 24 to 26°C and 75 to 80 per cent humidity are not available all through the year anywhere, but production can be effectively achieved if the ambient temperature in a region does not exceed 32°C or fall below 20°C at any time of the year. The parent-cocoon producing area should be nearby to ensure easier transport of the seed cocoons.

Building

The building should be spacious and cool, and should have separate rooms for cocoon storage, moth coupling, egg laying and acid treatment. In warmer regions, a two-storey building is preferred. Since the ground floor is cooler it can be used for cocoon storing and for *oviposition* (egg laying). The major part of the building will be utilized for cocoon production, but a room of 25m² floor area is required in which 50 000 bivoltine or 100 000 multivoltine cocoons can be stored.

The building should have well-ventilated, conveniently sized rooms for easy maintenance of temperature and humidity. As moths are sensitive to light, the building should have facilities to regulate light. It is very important to maintain cool temperatures in the room where coupling and oviposition takes place. In very warm areas it is better to get these rooms air-conditioned. An open space adjoining the building is necessary to prevent moth-scale dust from affecting the workers.

There should be a good, continuous water supply to the building. There should be an open yard on the premises for disinfecting the appliances and sun-drying the pierced cocoons (empty cocoon shells after the moths have emerged). A separate, well-ventilated, rodent-proof room or a shed situated away from the grainage should be made available for storing pierced cocoons.

Staff

The efficient management of the grainage by well-trained staff determines the quality and the viability of the egg production process. It is necessary for staff to have sufficient knowledge and skill in the processing and handling of eggs. They should be in a position to forecast seasonal demands and plan egg production accordingly.

Equipment

Production efficiency also depends on the good planning of the infrastructure in the grainage. Adequate equipment, preferably made locally, should be available. A list of appliances required for egg production in a grainage is provided at the end of this chapter.

Seed cocoon procurement

At least two or more distinctly different parent silkworm races are required for hybridization. The grainage is responsible for procuring and preserving seed cocoons in sufficient quantities to produce the required quantity of layings.

In sericulturally advanced countries, seed-cocoon production is well organized and farmers enlisted as seed growers work under mandatory regulations. In new areas, to ensure a continuous and adequate cocoon supply, the grainage should organize its own seed-cocoon procurement system by supplying pure seeds to selected farmers for rearing parent cocoons according to a well-planned schedule.

Quality assurance

As the grainage is committed to supplying good quality seeds, all measures for quality control should be strictly followed. The parent seed cocoon should be raised in hygienic conditions under scientific supervision. Moths should be regularly examined to make sure that the eggs are free from disease. The disease-free eggs should be further disinfected to avoid any surface contamination. The eggs should be incubated until they are supplied to the farmers.

Economics

There is a direct correlation between the quantity of eggs produced and profit. However, production cannot be increased indiscriminately without compromising quality. It is important to recognize that the production capacity is limited by the infrastructure, raw material availability and the demand for layings. Production is planned to make the unit economically viable.

Table 11 Cost of production of each laying

Overheads

❑ premises (rent or purchase)

❑ heat / light / water

❑ telephone / fax

❑ depreciation on equipment and interest on purchase loans

❑ consumable materials

❑ insurance

❑ postage and stationery

Direct costs

❑ wages

❑ raw material (parent cocoons)

❑ percentage of egg recovery✳ (25 to 35 per cent of female component of hybrid)

❑ transport

✳For example; from a batch of 100 moths, approximately 50 will be female. Perhaps only 40 of these female moths will lay eggs at all. Of these 40 egg laying females only 35 might have laid good eggs (12.5% of these layings will therefore be rejected). This means that there has been a 35% egg recovery.

The cost of producing one disease-free laying (DFL) can be worked out using the following formula:

$$\frac{\text{Cost of total female seed cocoons} + \text{Cost of total male seed cocoons} + \text{Other direct costs} + \text{Indirect costs}}{\text{Total number of DFLs prepared}}$$

Since the indirect costs are fixed, irrespective of the quantity of layings produced, and because seed cocoon costs fluctuate seasonally, the selling price should be regularly updated taking into account the total cost of production and the margin of profit expected. The pierced cocoons produced as a by-product are in great demand from the spun silk industry and revenue from selling them is an added income to the grainage.

SMALL GRAINAGE MODEL

Illustration 22 shows the basic layout for a small grainage suitable for the production of up to 50 000 DFLs per month. It has room for storing parent cocoons, coupling, and oviposition. It has sufficient space for processing eggs and it has an area for moth testing.

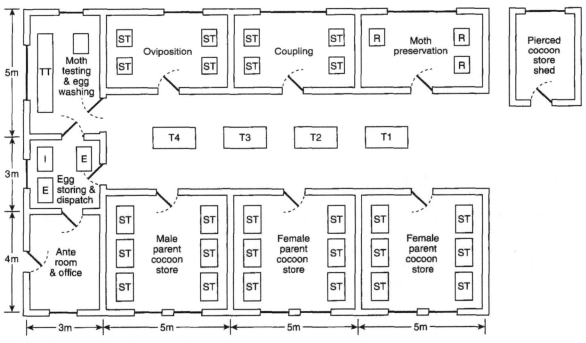

S – Racks for stacking cocoon trays
R – Refrigerator
T – Working tables
TT – Testing table
I – Incubator
E – Egg chamber

Illustration 22 Small grainage building layout for 50 000 DFL production capacity

Table 12 Grainage equipment for 50 000 DFL production capacity

Item	Quantity required
tray racks	20
wooden tray stands	8
wooden stools or benches	10
refrigerator	2
incubator	1
microscope	4
acid treatment bath	1
air conditioner	1
humidifiers	4
room heaters	4
centrifuge	1
moth-crushing equipment	10
hydrometers	2
hygrometers	4
thermometers	2
egg cabinet	1
seed cocoon bins	20
cellules	30000
sprayer	1
gas masks	2
washing trays	2
tinner	1
buckets, bowls, mugs and overall clothing	

PLANNING A CENTRALIZED SILK-REELING UNIT

There are several types of machine available for reeling raw silk. Each machine has a different level of mechanization. The simplest is the country charaka or *ITSEC* machine, and there is the cottage basin and the multi-end reeling machine. All are popular and supply different markets, depending on the type of raw silk they produce. The particular technology adopted by a reeler is determined by the main factors shown below.

❑   availability of raw material (cocoons)

❑   quality of cocoons produced in the area

❑   requirements of the weavers

❑   labour availability

❑   socio-economic conditions in the reeling areas

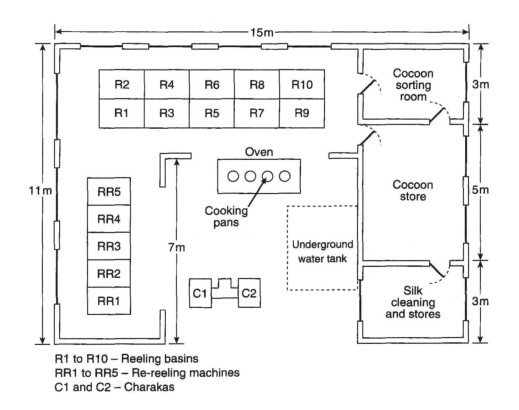

R1 to R10 – Reeling basins
RR1 to RR5 – Re-reeling machines
C1 and C2 – Charakas

Illustration 23 Layout for small reeling unit

Table 13 Economics of setting up a reeling unit

Overheads (indirect costs)

❑ premises / depreciation on building
❑ heat / light / water
❑ telephone / fax
❑ depreciation on equipment and interest on purchase loans
❑ consumable materials
❑ insurance
❑ postage and stationery
❑ interest on the cost of stock

Direct costs

❑ wages
❑ raw material (cocoons)
❑ fuel (firewood / husk / electricity)
❑ transport

The profitability of a reeling unit is dependent on the local cocoon prices and silk prices. A reeler should consider the costs listed in Table 13 before deciding the sale price for the silk. By considering the direct costs and the cost of overheads, the overall cost of production of 1kg raw silk yarn can be worked out as shown in the following equation.

$$\text{Cost of production of 1kg of raw silk yarn} = \frac{\text{Direct costs} + \text{Overheads}}{\text{Total kg raw silk produced}}$$

The profit margin should will be added to the cost of production to determine the selling price. The following information gives a basis for planning a unit. The example here is for 18/20 denier yarn production. Coarse yarn requires less time than finer yarn for production.

The term renditta describes the weight in kilogram of cocoons required to reel one kilogram of raw silk. The example shown in Table 14 is the renditta for bivoltine hybrid cocoons. Better quality cocoons, such as bivoltine, yield more silk giving a lower renditta. Low quality cocoons, chiefly multivoltine types, produce a higher renditta.

Renditta = number of kg cocoons to produce 1kg raw silk.

Table 14 Renditta for bivoltine hybrid cocoons

Reeling machine	Cocoons (g)*	Renditta	Raw silk produced (g)
charaka	1500	10	150
cottage basin	900	11	82
multi-end	1000	11	91
semi-automatic	1000	12	83
automatic	1000	12	83

Weight of cocoons reeled in one day (eight hours)

6. EQUIPMENT SUPPLIERS

Japan

Sericulture equipment, reeling devices, cocoon deflossing machines

Sobajuma Co. Ltd
No. 8-12, 1 Chome, Noritake
Nakamura - Ku Nagoya

Cocoon boiling machines, re-reeling machines, testing machines

Toyo Sangyo Consulting Inc.
No. 608, Laurel Nagalacho Building
No. 17-5, 2 Chome, Nagata-Cho
Chiyoda - Ku Tokyo - 100

Automatic reeling machines, rearing equipment, grainage equipment

Nissan Company Limited
Textile Machine Division
3-1, 5 Chome Shimorenjaku
Mitaka, Tokyo

India

Silk reeling machines, warping machines, throwing units

Alltex Textile Engineers
Lalbagh Fort Road
Bangalore 560 004

Aryan Rural Industries
Kanakapura
Bangalore District

Star Engineering Works
26 Industrial Area
Bangalore 560 003

Gajalakshmi Engineering Works
Magadi Road
Bangalore 560 023

National Textile Engineering Works
2 Basappa cross road
Shanthinagar
Bangalore 560 027

7. SOURCES OF FURTHER INFORMATION

INSTITUTIONS AND ORGANIZATIONS

Central Sericulture
Research and Training Institute
Manandavadi Road
Srirampuram
Mysore - 570 008
India

Central Silk Board
United Mansion
39, Mahatma Gandhi Road
Bangalore - 560 001
India

Central Silk Technological Research Institute
BTM Layout
Madivala
Bangalore
India

EDA Rural Systems
107 Qutab Plaza
DLF Qutab Enclave Phase - 1
Gurgaon 122002
India

Food and Agriculture Organisation
of the United Nations
Via delle Terme Di Caracalla
00100 Rome
Italy

Institute of Silkworm Genetics and Breeding
Dainippon Raw Silk Foundation
1054, Likura Amy Machi
Ibaraki-ken. 30004
Japan

Intermediate Technology
Myson House, Railway Terrace
Rugby, CV21 3HT
United Kingdom

International Silk Association
20 rue Joseph, Serlin
69001, Lyons
France

Karnataka State Sericulture
Research and Development Institute
Thalaghattapura
Kanakapura Road
Bangalore
India

Sericulture Experiment Station
61, Seodun-Dong Suwon
Republic of Korea. 170

Seri Tech Associates
No.9 11th Cross
Sampangiram Nagar
Bangalore 560027
India

Toyo Sangyo Consulting
2-7-5, Nagoda Cho
Chiyoda-Ku
Tokyo
Japan

Worldwide Butterflies Limited
Compton House
Sherborne
Dorset DT9 4QN
United Kingdom

LITERATURE

Handbook of Silkworm Rearing, Fuji Publishing Company Limited, Tokyo, Japan. 1972

China: Sericulture, FAO Agricultural Services Bulletin 42, FAO, Rome, Italy. 1980

Mulberry Cultivation, FAO Agricultural Services Bulletin 73/1, FAO, Rome, Italy. 1988

Silkworm Rearing, FAO Agricultural Services Bulletin 73/2, FAO, Rome, Italy. 1988

Silkworm Egg Production, FAO Agricultural Services Bulletin 73/3, FAO, Rome, Italy. 1989

Silkworm Rearing and Disease, Yenimura, Mysore Silk Reelers Association, Bangalore, India.

Sericulture in Japan, CSB Bulletin, Central Silk Board, Bangalore, India.

Silk in India, Statistical Biennial, Central Silk Board, India.

Silk from Grub to Glamour, N.Nanavathy, Paramount Publishing House, Bombay, India.

Economics of the Silk Industry, R.C.Rawley, P.S.King & Sons Limited, London, UK.

Handbook of Practical Sericulture, S.R.Ullal and M.N.Narasimhanna, India.

Textbook of Tropical Sericulture, S.R.Ullal and M.N.Narasimhanna, India.

Appropriate Sericulture Techniques, M.S.Jolly. CSR+TI, Mysore, India.

The Queen of Textiles, Nina Hyde, National Geographic, USA. 1984

The Silk Book, The Silk and Rayon Users' Association Incorporated, London, UK.

Testing of Yarns and Fabrics, N.Eyre, Emmotl and Co. Ltd.

Silk Biology, Chemistry and Technology, Paolo Carboni, Chapman & Hall, London, UK.

The Development of Indian Silk, Sanjay Sinha, Intermediate Technology Publications, London, and Oxford & IBH Publishing Co Pvt Ltd, New Delhi, India.

The Romance of Textiles, Ethel Lewis, Macmillan Co. of Canada, Toronto, Canada.

Synthesised Science of Sericulture, Central Silk Board, Bangalore, India.

Sericulture - Training Manual for Women, Dr Neeru Saluja, National Institute of Public Co-operation and Child Development, New Delhi, India. 1989

Charaka Silk Reeling: Technical Options, Edward Thomas, Intermediate Technology Development Group, UK in collaboration with Economic Rural Systems, Gurgaon, Near New Delhi, India. 1994

The Story of Silk, Dr John Feltwell, Alan Sutton Publishing Limited, Far Thrupp, Stroud, UK. 1990

Fabric Manufacture: A handbook, Alan Newton, Small-scale textiles series, Intermediate Technology Publications, UK. 1993

The Illustrated Dictionary of Fabrics, Martin Hardingham, Studio Vista, London, UK. 1978

APPENDIX 1
REELING DEVICES

The principal function of a reeling device is to unravel filaments drawn from several cocoons to form a single yarn. Crude and cumbersome reeling devices of the earlier days have now given rise to many sophisticated and highly-developed reeling machines. Manually operated reeling machines are also still in vogue. The process of reeling is being practised in different countries using several reeling devices, some of which are described here.

COUNTRY CHARAKA

The country charaka (Illustration 24), is a traditional reeling machine which is made and used in India. It is a manually-operated reeling device extensively used in the rural reeling centres. It is locally built using the available material and technology. The country charaka does not conform to any specific design or standard. The charaka is constructed with a mud platform, a simple manually-controlled reel which is usually made of wood, and a simple distributor system.

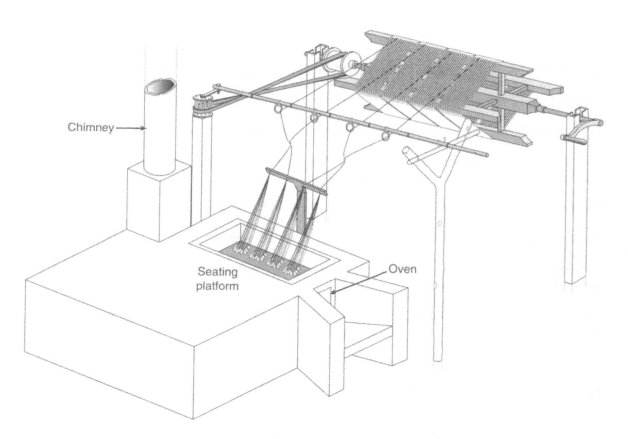

Illustration 24 Typical country charaka

Platform

This is a rectangular block with a built-in fireplace on which a vessel made of aluminium or copper is placed. It is used for cooking cocoons and reeling. The fireplace is built with a chimney and an oven for heating the pan from below. The vessel is 50cm in diameter at its mouth which reduces to 20cm at the bottom. The platform is wide enough to provide space for the reeler to squat. Above the vessel, overhanging it, is a metal strip which is referred to as tarapatti. It has four holes spaced evenly through which filaments are passed while reeling. These function as the thread guides.

The reel

In a charaka, a single long reel made of thick, seasoned wood is used. The circumference of the reel varies from 150 to 225cm. The reel is mounted on wooden posts or metal frames in such a way that it rotates freely on its horizontal axis. A handle situated at one end of the reel is used for turning the reel.

The distributor

The distributor consists of a simple, crudely-made metal or wooden wheel fixed in front of the reel and it revolves on its vertical axis. A wooden traverse rod is attached to the wheel at an eccentric point. A loop of string is tied around the wheel, and the reel shaft transmits the drive from the reel to the distributor wheel, which results in the sideways movement of the distributor rod. The traverse rod has four loops of wire or rings along its length serving as thread guides distributing the silk on the reel.

Between the *tarapatti* and the distributor, the threads emerging from two thread guides located next to each other are entwined to form a *chambon croissure* before they are wound on the reel. The *chambon croissure* is characteristic of charaka reeling.

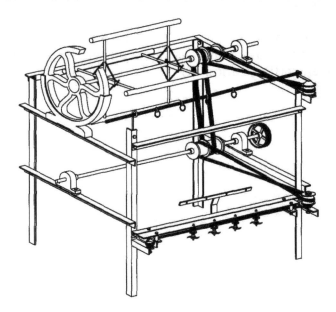

Illustration 25 ITSEC improved charaka

Intermediate Technology, UK, has over the past few years developed a series of three improved charakas in South India, in co-operation with the Central Silk Board of India. The indigenous charaka has been modified, the three improved designs, one manual and two power operated, are known as the *ITSEC* charakas (Illustration 25). The *ITSEC* charakas contain modifications which provide a better quality raw silk and an increased output.

DOMESTIC BASIN

This is an improved version of the country charaka. The domestic basin utilizes better-quality cocoons more efficiently and produces better and finer silk. It has three very distinct but functionally connected parts: the cooking unit, the reel bench and the reel frame. Silk reeled on a domestic basin does not require re-reeling, as the silk is reeled directly on to a standard reel.

Cooking unit

When using the domestic basin the cooking unit is isolated from the main reeling machine. It consists of a mud or cement platform on which several cooking pans are fixed in a row. The oven is well constructed with gratings, ash pits, a chimney and an inlet for fuel. The cocoons are cooked in the pan first, brushed and then transferred to the reeling basin for reeling.

Reel bench

Each reeling unit normally consists of five to six reeling basins. The reeling bench is usually made of cast iron supporting frames on which the basins are fitted in a row. The basins are made of either stainless steel, copper or prefabricated cement. They are shallow rectangular troughs designed to reel four to six ends of silk. The croissure system adopted is of the tavelette type. The croissure pulleys are fitted on a frame which is fixed vertically to the bench. Porcelain buttons with small holes in the centre are fitted on the frame. These function both as thread guides and as slub catchers. In some machines each porcelain button is replaced with a *jettabout*, which improves cocoon casting efficiency (Illustration 26).

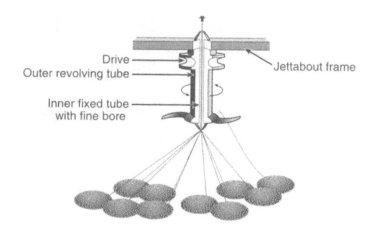

Illustration 26 The jettabout

50

Reel frame

The reel frame is fitted towards the back of the reeler about 100cm away from the reel bench at a height of 150 to 170cm from the ground. This provides sufficient distance for the thread to dry before it winds on to the reel. The reels are of standard size, as prescribed by international standards. They are turned by drive wheels which are fitted on a common transmission shaft. One end of the transmission shaft also drives the traverse mechanism through a series of gears. At the other end, a hand-operated braking mechanism is provided to stop reels individually whenever necessary.

COTTAGE BASIN

The cottage basin (Illustration 27) is a much simplified version of the multi-end reeling machine. The cottage basin is similar to the domestic basin with a reel bench, *croissure* and reel frame with small reels, which are set in the reel frame directly above the croissure frame, but with a separate cooking system. Re-reeling is necessary from the small reels.

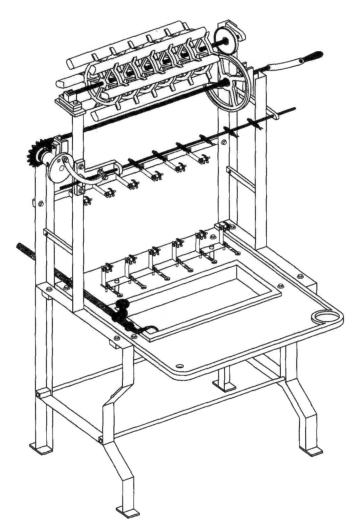

Illustration 27 Cottage basin reeling machine

MULTI-END REELING MACHINE

The power-driven multi-end reeling machine is popular in those countries where sericulture and silk production are more advanced. This machine is used for processing superior grade cocoons. Unlike the cottage basin it is slower and more suited to uniform reeling. It has more than six ends per basin. The multi-end reeling device has been developed for producing high quality raw silk and for high productivity.

The multi-end reeling machine is compact and it is designed to make the reeler's work less strenuous. The height of the reeling bench is so adjusted that it is convenient for the reeler to either sit or stand while reeling. The basin is a long rectangular trough, accommodating several smaller compartments for keeping reserve cocoons, for casting unreeled cocoons, and for waste pupae. The thread guide frames are fitted with *jettabouts* for easier cocoon casting.

The reeling spools are 60 to 75cm in circumference, made of either light metal, hard wood or plastic. Each reel on the shaft has an individual brake, operated by a lever mechanism which stops the reel whenever there is a change in the tension of the reeling yarn. All the reels can also be stopped simultaneously if necessary. The slow speed of the machine, in addition to the introduction of the *jettabout* and the individual brake system contribute to easier reeling and have also enhanced the quality of silk produced.

AUTOMATIC REELING MACHINE

The fully automatic, power driven, reeling machine has been developed in Japan. This hi-tech machine, often having as many as one or two hundred reels per machine, is used in countries where there is a dearth of skilled labour.

As the name implies, various operations like cocoon cooking, brushing, picking, feeding and reeling are mechanized in the automatic reeling machine. The machines are also designed to regulate thread size. There are two systems for controlling the size of thread. One is the *constant size* type and the other is the *fixed number* type. The size of thread is revised either by monitoring the friction between the reeling thread and the size control device or by monitoring the stretching stress of the yarn, which changes according to the size of the yarn.

This very sophisticated machine is able to supply cocoons automatically to the cocoon feed, it automatically finds the reeling ends of silk from the cocoons, it gathers dropped cocoons, removes pupae, cleans the basins and regulates the reeling water temperature. The quality of raw material is important for the efficient functioning of the machine. For reeling 20/22 denier silk, the cocoon characteristics are important and should meet the minimum requirements, shown below, to give an even thickness 20.5 denier silk filament.

❑   length of combined filaments   1000m

❑   reelability   70%

❑   length of unbroken cocoon filament   700m

❑   thickness of cocoon filament   2.56 denier

APPENDIX 2
GLOSSARY

The phonetic spelling of *silk* in the local language of each country where silk is produced is as follows:

see	**Chinese**
silk	**English**
soie	**French**
scide	**German**
serikon	**Greek**
seta	**Italian**
sir	**Korean**
sericum	**Latin**
seda	**Spanish**
sheolk	**Russian**

all silk	a yarn or fabric which contains no other textile fibre other than silk.
bave	the raw silk filament exuded by the silkworm in the form of two brins. See also *brin*.
bivoltine	a silkmoth which produces two generations per year and lays hibernating and non-hibernating eggs.
bolting cloth	plain woven sheer silk fabric used for sifting. From the French word *blutage*. Also known as miller's gauze.
book	sixteen to twenty skeins of raw silk compactly packed into a parcel weighing from 2 to 2.5kg.
brin	two brins are exuded from the head of the silkworm to form the bave or silk filament.
chambon	the chambon croissure (*French*) is composed of two groups of filament which cross between the cocoon and the distributor on a reeling machine. A similar *Italian* device is called a *tavelette*.
chawki	newly hatched silkworm.
chiffon	a very light, diaphanous fabric. Both warp and weft yarns used are highly twisted *crêpe*. Unlike in crêpe de Chine, the weft yarn is either **S** or **Z** twist. The characteristic wrinkles in the finished fabric are created by the weft yarns being pulled in one direction. From the French word literally meaning a *rag*.
compenzine	made of two silk yarns of which one is single twisted, the other is untwisted yarn. When they are twisted together, the resultant yarn crinkles up along its length giving a knobbly appearance.
crêpe de Chine	soft, thin, opaque and lightweight fabric. Woven with highly twisted weft threads and untwisted warp threads. Alternate picks are of opposite twists resulting in a crimpy appearance on the fabric.

denier	part of the direct, fixed-length, count system to determine the size of a filament yarn. The number of grams per 9000 metres.
eri silk	eri silkworms thrive on castor oil leaf to produce their cocoons which are usually white but often golden in colour.
fibroin	a protein chemical substance which is not soluble in water and which forms the core of the silk filament.
floss	the loose silk round the cocoon which is retained, before reeling is started, and used in the production of spun silk. Also the name given to some low twist silk embroidery yarns.
foulard	a 2 + 2 twill soft, lustrous, silk fabric originally woven in India.
grenadine	grenadine is a tightly twisted yarn in which two or three single twisted strands are plied and double twisted in the opposite direction more tightly than organzine giving it extra strength in weaving and a dull appearance to the fabric.
grosgrain	a silk fabric with pronounced ribs across a heavy cloth. From the French *gros*, meaning large and *grain*, meaning cord.
georgette	very thin, transparent or semi-transparent fabric, which is more grainy than crepe. This quality is the result of highly twisted warp and weft threads.
grainage	a building or set of buildings in which disease-free silkworm eggs (DFL) are produced under strict, hygienic conditions. From the word grain meaning seed.
habutai	a lightweight Japanese silk fabric. Sometimes refered to as Jap cloth.
jacquard	a device, invented by Joseph Marie Jacquard between 1801 and 1810, which is attached to a loom to weave pictures and designs. The mechanism is operated by either a punched card system or computer in order to select individual warp threads.
jettabout	an attachment to the silk reeling machine which simplifies the process of taking in fresh filaments from the cocoon during reeling.
muga silk	muga silkworms, found in Assam, northern India, belonging to the same genus as tussah, live on leaves from hance (*liquidambar formosana*) produce a fine, strong, golden coloured silk.
mutka	also spelled matka, a silk cloth woven from handspun mulberry silk waste.
nett silk	filament silk drawn off the cocoon as a continuous thread.
organzine	a strong silk yarn made from high quality silk. Single silk yarn is twisted and then doubled. The compound thread is twisted once again in the opposite direction resulting in 350 to 1300 tpm(twists per metre). Organzine is mostly used for warp.
multivoltine	also polyvoltine; a silkmoth variety which produces several generations per year and lays only non-hibernating eggs.
pongee	the Chinese word *pen-chi* means handwoven or woven at home. Other types of pongee are: shantung, hohan, antung and ninghai. The warp is finer than the weft which is usually a dupion yarn often mistaken for so-called wild silk because of its creamy colour.

pure silk	a silk yarn or fabric which contains no metallic or other weighting agents except those essential ones used in dyeing.
pure dye silk	no weighting of any kind is used even during dyeing.
raw silk	silk filaments which have been reeled but are unprocessed and still containing *serecin*.
renditta	the term used to determine the weight in kilograms of cocoons to reel 1kg of raw silk.
schappe silk	spun silk woven fabric which has been degummed by fermentation.
serecin	the protein liquid, also known as *gum*, which coats the silk as it is exuded by the silkworm.
shot silk	like taffeta, woven with one colour in the warp and contrasting colour in the weft.
spun silk	yarn spun from waste silk which is processed and spun like cotton.
stone washed	silk fabric which has been washed with sand to create a soft handle to the fabric.
swift	a simple metal or wooden frame normally 36 inches or 1 metre in circumference and supported on a stand, rotated by hand or motor. A simple device on which to wind yarn.
tarapatti	a metal strip pierced with four small holes through which the silk passes during reeling. The strip of metal is parallel to the floor above the vessel containing the cocoons and hot water.
tasar	see *tussah*.
tram	tram is medium twisted thread formed by twisting 2 to 3 silk yarns together with low twists of 100 to 150 tpm. It is moderately strong, soft and has a good handle (feel) and is mostly used as weft.
tussah	from the Hindu word *tasar* which is derived from the Sanscrit *tasara* meaning shuttle. A variety of hard silk from the cocoon produced by the tussah silkworm, from the genus *antheraea*, which spins its cocoon from the branches of the *terminalia* tree, the leaves of which it eats. Often referred to as wild silk.
univoltine	silkworm variety, native only to temperate regions, and which produces one generation of only dormant eggs per year.
waste silk	the floss on the outside of the cocoon before it is reeled and the final layer of silk around the pupa after reeling. Also any bad unreelable cocoons. This waste silk can be processed and spun like cotton.
wild silk	see *tussah*.